GREEN LEVIATHAN

Federalism Studies

Series Editor: Søren Dosenrode

The end of the Cold War profoundly altered the dynamics between and within the various states in Europe and the rest of the World, resulting in a resurgence of interest in the concept of federalism. This shift in balance has been further fuelled by the increase in the number of conflicts arising from the disaffection of the diverse ethnic or religious minorities residing within these states (e.g. Sudan, Iraq). Furthermore, globalization is forcing governments not only to work together, but also to reconsider their internal roles as guarantors of economic growth, with regions playing the major part.

It is the aim of the series to look at federal or federated states in historical,theoretical and comparative contexts. Thus it will be possible to build a common framework for the constructive analysis of federalism on the meta-level, and this in turn will enable us to identify and define federal tradition traditions, and develop the theoretical.

This unique and ground-breaking new series aims to promote a complete and indepth understanding of federalism by collectively bringing together the work of political scientists, lawyers, historians, economists, sociologists and anthropologists, and with this in mind, contributions are welcomed from authors in all of these disciplines. But whereas the federal approach is the crank of the series, it does not mean that contributions must adhere to the federal approach; critical contributions are welcome too.

Also in the series

Green Leviathan
The Case for a Federal Role in Environmental Policy

INGER WEIBUST
Carleton University, Canada

Routledge
Taylor & Francis Group

LONDON AND NEW YORK

First published 2009 by Ashgate Publishing

2 Park Square, Milton Park, Abingdon, Oxfordshire OX14 4RN
52 Vanderbilt Avenue, New York, NY 10017

Routledge is an imprint of the Taylor & Francis Group, an informa business

First issued in paperback 2020

British Library Cataloguing in Publication Data
Weibust, Inger.
    Green leviathan : the case for a federal role in
    environmental policy.--(Federalism studies)
    1. Environmental policy--Case studies. 2. Local government
    and environmental policy--Case studies. 3. Federal
    government--Case studies. 4. Central-local government
    relations--Case studies.
    I. Title II. Series
    363.7'056-dc22

Library of Congress Cataloging-in-Publication Data
Weibust, Inger.
    Green leviathan : the case for a federal role in environmental policy / by Inger Weibust.
       p. cm. -- (Federalism studies)
    Includes bibliographical references and index.
    ISBN 978-0-7546-7729-1 (hardcover) -- ISBN 978-0-7546-9446-5 (ebook)
    1. Environmental policy. 2. Federal government. I. Title.

    GE170.W445 2009
    363.7'0561--dc22

                                                                2009003420

ISBN 13: 978-0-7546-7729-1 (hbk)
ISBN 13: 978-0-367-60558-2 (pbk)

Contents

List of Figures

List of Tables

Acknowledgements

This book was made possible by the support and encouragement of many people. In developing this project, I benefited greatly from Kenneth Oye's intellectual guidance and his enthusiastic support. Suzanne Berger's simple yet penetrating questions were very important in shaping the final design of the project. Sharon Sutherland, Jo Solet and Kim Hayes were excellent listeners and offered many constructive suggestions for completing the project.

I also benefited from discussions with and feedback from friends. Among those who offered food for thought are Loren King, Tamar Gutner, Gunnar Trumbull, David Reiner, P. R. Goldstone, David Mussington, Dale Murphy and Katharina Holzinger.

I would like to acknowledge the financial support I received which helped fund my fieldwork: the MIT Center for International Studies, Harvard University's Center for European Studies, the V. Kann Rasmussen Foundation, the Alliance for Global Sustainability, the MIT-Volvo Award for Environmental Research and the Social Sciences and Humanities Research Council of Canada.

I received help and logistical support from many people, in many places in the course of carrying out the research. Among the most important are the kind souls who opened their (sometimes tiny) apartments to me: Michel Goyer, Marga Gomez-Reino and Douglas Pipe. It would be impossible to thank all the people in Europe and North America who took the time to speak with me. Among these helpful people, John Wolf of the Kommission Reinhaltung der Luft stands out for taking such an interest in my research. I also benefited greatly from the helpfulness of the staff at CITEPA in Paris and the International Union of Air Pollution Prevention Associations in Brighton. My research benefited enormously from the resources of the Harvard University libraries and the MIT Interlibrary Loan staff, who procured all but one of the hundreds of requests I made.

Finally, I could never have completed this project without the love and steadfast support of my parents, Hjørdis and Thorstein. They stood by me wholeheartedly through this long process and some very difficult times. I hope I can repay their generosity. My extended family also kept me in their thoughts and prayers. A special thanks goes to my husband, David Mendeloff. He has been my rock, who has given his time, thoughtful comments and encouragement. He is also a devoted father to our sons, Elias and Jonas, who despite knowing nothing about this book, have never known life without this project in it.

List of Abbreviations

ACE	Army Corps of Engineeers
AIT	Agreement on Internal Trade
BIGA	Bundesamt für Industrie, Gewerbe und Arbeit
BOD	Biological Oxygen Demand
CCME	Canadian Council of Ministers of the Environment
CEC	Commission for Environmental Co-operation
CEPA	*Canadian Environmental Protection Act*
COD	Chemical Oxygen Demand
CWS	Canada–Wide Standards
EEA	European Environment Agency
EPA	Environmental Protection Agency (US)
GHG	Greenhouse Gases
IGA	Intergovernmental Arrangement
IJC	International Joint Commission
ISSP	International Social Survey Programme
MSW	municipal solid waste
NO_x	nitrogen oxides
NPRI	National Pollutant Release Inventory
OECD	Organization for Economic Cooperation and Development
OTAG	Ozone Transport Assessment Group
POGG	Peace, Order and Good Government
QMV	Qualified Majority Voting
SF	Swiss Francs
SO_x	sulphur oxides
TRI	Toxics Release Inventory
TSS	Total Suspended Solids
TVA	Tennessee Valley Authority
VOCs	Volatile Organic Compounds

For my family

Chapter 1

Introduction

Does it matter who does what? In a federal system, does it matter whether environmental standards are set by the federal government or by states? Many environmentalists favour centralized environmental policymaking at the federal or European Union level. Should they? Furthermore, can state or provincial governments, acting cooperatively, achieve the same results as action at the national level?

This book addresses these questions by examining environmental regulation over several decades in the United States, Switzerland, Canada and the European Union. It demonstrates that centralized policymaking produces more stringent environmental policies than decentralized policymaking. In all the cases where policymaking became more centralized, environmental policy subsequently became more stringent. Where formal and informal cooperation were tried, they proved less effective at improving environmental quality than the national standard setting that superseded them.

According to economic theory, it is possible for groups to provide themselves with public goods through voluntary, noncentralized mechanisms (Coase 1960; Buchanan 1968). However, there is little evidence that this happens if the public goods are clean air and water. Federal systems, with incentives to prevent defection and free riding, would seem to be conducive to this kind of public goods provision. In a federation, national courts are available to enforce contracts and the availability of fiscal transfers makes side payments possible. Yet even here, we do not observe effective cooperation for *particular kinds* of public goods, including environmental standards, the focus of this book.

Through the study of environmental regulations in three federations, this book develops two distinct arguments, by testing four hypotheses. The three federations are those a handbook on federalism characterizes the world's three classic federations: the United States, Canada and Switzerland (Elazar 1991). The findings from these cases are then applied to the European Union, a federal system but not a federation. The three federations represent critical cases, selected to be as conducive as possible to effective noncentralized or cooperative environmental policymaking. If we do not observe effective noncentralized or cooperative policymaking in the three federations, we are unlikely to observe it anywhere. To that end, the United States, Canada and Switzerland were selected as examples of wealthy countries, showing high levels of concern about the environment.

The Two Arguments

This study offers two distinct arguments, which have implications for two debates in political science, public policy and international relations theory. The first debate concerns the best way to address environmental problems at the national level. The second debate concerns the feasibility of cooperative solutions to environmental problems. The hypotheses which the study seeks to disprove are summarized in Table 1.1.

The first argument is that a noncentralized system of environmental policymaking will produce lower levels of environmental protection than a centralized one, *ceteris paribus*. This is true even when the impact of the pollution is primarily local and there are no transboundary environmental spillovers . Fear of competitive economic disadvantage inhibits action on pollution control. Although standards are less stringent in a noncentralized system, races to the bottom are rare. They are rare because, at equilibrium, standards are already rather lax. In the absence of referenda or popular ballot initiatives, it is difficult for the electorate of a noncentralized federation to raise environmental standards or to centralize standard setting, even if this is what the majority wants. A transition to centralized governance results in increased stringency in standards.

Table 1.1 Hypotheses and predictions

Hypothesis	Prediction
Hypotheses on centralization and policymaking	
H_1: National governments are no better able to make environmental policy than subnational governments.	There should be no systematic relationship between centralization of environmental policymaking and the level of environmental protection.
H_2: Environmental policymaking by lower levels of jurisdiction is more likely to be tailored to local environmental conditions than centrally set policies.	Environmental policies set at lower levels of jurisdiction should show diversity, reflecting diversity of local conditions.
H_3: Lower levels of jurisdiction better reflect voters' preferences.	In referenda and opinion polls, voters will prefer less centralized environmental policymaking.
Hypothesis on cooperation in federal systems	
H_4: In the presence of favourable contracting conditions to deter defection, subnational governments will achieve effective cooperation on environmental problems.	In federal systems with high levels of concern about the environment, we should find many examples of effective cooperation.

Second, even in the presence of strong legal systems for contract enforcement, environmental cooperation is rare and rarely effective in federations. Even when subnational governments have the recourse to external enforcement for their environmental agreements, they do not use it. To promote cooperation, it is not enough to assuage the fear of defection. Fears of lost competitive advantage remain. Nor is the availability of side payments sufficient to generate cooperation. Even on rivers or lakes, where those cooperating can be assured of capturing most or all of the benefits of pollution abatement, agreements are infrequent. Governments hesitate to participate because they fear giving an economic advantage to jurisdictions outside the watershed. Transboundary externalities and concerns about interjurisdictional economic competition mean that optimal environmental policy can only be attained though the imposition of floor standards by a central authority.

Environmental Federalism: What are the Consequences of Noncentralized Environmental Policymaking?

Arguments in favour of decentralized environmental governance come from a variety of sources. In North America, advocates of environmental federalism claim that the federal role in environmental policy should be drastically scaled back. They portray a federal government run amok, trampling on local prerogatives and causing overregulation (Anderson and Hill 1997). Scholars of Multilevel Governance describe a world in which 'centralized authority has given way to new forms of governing'(Marks and Hooghe 2004, 15). Furthermore, not only has governance become multi–level, it *should be* multi–level (Hooghe and Marks 2003, 233). Thus, decentralized environmental governance is optimal environmental governance.

This book claims that centralizing environmental policy produces higher levels of environmental protection, other things being equal. Environmental regulation differs from other forms of economic regulation such as rate regulation because it concerns externalities which often cross borders. Whereas competition can replace many forms of economic regulation, it is not a substitute for environmental regulation. Competition between jurisdictions or between firms makes environmental externalities worse, because the cost of production does not reflect the true social cost of the product. Richard Revesz and others argue that there are economic and environmental benefits to be gained from interjurisdictional environmental competition (Butler and Macey 1996). In the absence of legal provisions for emissions taxes or regulations, producers pay nothing for resources such as clean air. 'Free' resources will be overused because their price does not accurately reflect their value to society. The optimal solution to the problem of environmental regulation in a federal system is the central imposition of minimum standards or taxes.

In a federation, subnational units are unable to address these issues adequately, individually or collectively. On an individual basis, they are reluctant to regulate pollution because they are very concerned about the potential loss of business and investment to other jurisdictions. This is true even when the pollution does not have transboundary impacts. According to the economist's account of environmental regulation, local regulation should be optimal, in the absence of transboundary pollution spillovers (Oates and Schwab 1988). The hypothesis to be falsified is: *national governments are no better able to make environmental policy than subnational governments.* The hypothesis would be confirmed if we find no systematic relationship between centralization and level of environmental protection or efficacy of environmental policy.

Another argument in favour of decentralization is that states or provinces are better able to make environmental policy, because they are smaller than nation–states. Lower levels of jurisdiction are supposed to have more accurate knowledge about local conditions, local problems and local preferences. On the basis of better knowledge of local conditions and better representation of preferences, it is argued that lower levels of jurisdiction make policies appropriately tailored to local conditions, whereas higher levels of government will tend to impose a one–size–fits–all solution. Stated as a hypothesis, the claim is: *environmental policymaking by lower levels of jurisdiction is more likely to be tailored to local environmental conditions than centrally set policies.* This hypothesis would be confirmed by a finding of diverse policy responses, targeted to local problems, across regions in subnational environmental policies.

One of the strongest arguments in favour decentralization is the claim that smaller jurisdictions are more representative. Smaller, more numerous jurisdictions should permit better preference aggregation, such that a larger number of people have their preferences satisfied. However, there are limitations due to technology or collective action problems. Technological limits, such as externalities or economies of scale, are one type. For example, nuclear weapons policy cannot be made on a city by city basis.

A second type of limitation is collective action problems, such as the Prisoner's Dilemma. If a problem can be characterized as a Prisoner's Dilemma, then lower level decision–making is less optimal than a centralized decision. The hypothesis here is: *lower levels of jurisdiction better reflect voters' preferences than higher levels of government.* This hypothesis would be confirmed if referenda and opinion polls show popular support for decentralized environmental policymaking.

Cooperation Within Federations and the International System

Cooperation on the environment poses an even greater challenge at the international level than within federations. With few exceptions, the international relations literature has not taken the experiences of federal systems into account in generating or testing theory.[1] Theories about international environmental cooperation are derived from formal models or observations about international agreements. Both abstract models of cooperation and empirically based theories should have some predictive validity for cooperation in federal systems. If these theories are inconsistent with the results observed in federal systems, they might require revision. If research shows that subnational governments in federal systems are rarely able to address environmental problems collectively, then, according to leading theories of international cooperation, the prognosis for effective action at the international level is poor.

Most scholars of international environmental affairs believe that institutional design is very important in addressing the world's environmental problems. Institutions such as international regimes are thought to reduce transaction costs and make coordinated action possible. 'Three C's': a high level of *concern*, a favourable *contracting* environment and *capacity* have been identified as necessary preconditions for effective international environmental institutions (Keohane et al. 1993, 11).

A federation should be superior to an international regime in providing the facilitating conditions for cooperation. If an international institution is able to provide favourable conditions for contracting, through side payments, repeated interaction, monitoring and enforcement, then *a fortiori* a federal system should be even better able to facilitate binding agreements by the federal submits. Stated as a hypothesis: *subnational governments will achieve effective cooperation on environmental problems, in the presence of favourable contracting conditions to deter defection.* This hypothesis would be confirmed if we find numerous examples of effective cooperation in federal systems with high levels of concern about the environment.

Case Selection

The four cases have some common features but also have other characteristics that are case specific. All of the cases have the capacity for informal cooperation and two (the US and Switzerland) have constitutional provisions which permit formal, legally binding cooperation among states or cantons. The US is distinctive because it has the most extensive experience of formalized cooperation on the environment, more than 70 years. Switzerland is the only case where the public can effectuate a redistribution of constitutional powers, through the referendum mechanism. The Swiss have a long history of referenda on environmental issues, dating back to the nineteenth century. Canada is the only case that has not undergone systematic and far–reaching centralization of environmental policymaking. Thus Canada serves

---

1 Daniel Deudney (1996) examined the evolution of the United States to explore concepts of sovereignty and security, which he argued had implications for the European Union.

as a counterfactual: what would environmental policymaking look like in the absence of the centralization which occurred in the 1970s and 1980s in most federal systems. The European Union is distinctive because although it makes decisions through intergovernmental cooperation, enforcement provisions lie outside any particular environmental regulation and are not subject to negotiation.

Structure of the Book

The book is divided into nine chapters. Chapter 2 examines the claims made in favour of decentralized governance. Chapter 3 discusses evidence about the incidence and effects of interjurisdictional competition, including races to the bottom. Chapter 4 reviews the literature on cooperation between governments. Chapter 5 presents the American case, focusing on the American experience with state cooperation on the environment. Chapter 6 presents the Swiss case, focusing on the increasing centralization of environmental policy brought about by a series of referenda. Chapter 7 presents the Canadian case, contrasting Canadian and American data on opinion polls and environmental protection. Chapter 8 applies the findings from the first three cases to the European Union and discusses the implications of unanimous decision–making. Chapter 9 presents the conclusions of the book. It summarizes the findings from the cases, presents comparative analysis across the cases and ends by identifying areas requiring further research.

Chapter 2

Examining the Case for Decentralized Policymaking

Decentralizing policymaking is all the rage, from the US Congress to the European Union with its goal of subsidiarity. Of late, belief in the benefits of decentralized governance has grown so strong that the burden of proof now lies almost entirely with those favouring centralization. In the US, decentralization has advocates on both the right and the left. Those on the right see the decentralization of government as a way to harness competitive forces to ensure more efficient government and less of it. Those on the left see more decentralized policymaking as a means to empower communities and make policy more responsive to the little guy.

Environmental Federalism

Environmental policymaking has not been insulated from this debate. An entire cottage industry, termed 'environmental federalism' has grown up around this issue (for example, Anderson and Hill 1997). The focus here is the sovereignty of state governments in environmental policymaking; administrative decentralization to regional offices of the Environmental Protection Agency is not the goal (Rubin and Feeley 1994, 41). Writings in favour environmental federalism have generally emerged from conservative think tanks such as the American Enterprise Institute (Butler and Macey 1996). The liberal and left–wing environmentalist communities are of two minds on the issue. There is widespread suspicion that turning environmental policy over to the states would result in competition in laxity and lower levels of environmental protection. On the other hand, the violent protests surrounding the 1999 meeting of the World Trade Organization in Seattle point to fears that supranational organizations will force Americans to adopt weaker standards or will constrain their ability to set stringent standards. As a rule, environmentalists would like to maximize the capacity for governments at all levels to set higher standards.

The chapter has three sections. The first discusses the economic concept of optimal jurisdiction and why state governments are often assumed to represent the optimal level of jurisdiction for environmental policy. The second section focuses on the claims made for decentralization and the empirical evidence, if any, for those claims. These claims can be grouped into six types of arguments: information, preference aggregation, representation, effectiveness, innovation and legitimacy. The third section summarizes conclusions from the chapter.

Optimal Jurisdiction

Even if one accepts the case for decentralization, this case does not necessarily establish the states as the optimal level of jurisdiction, particularly for environmental decision–making. Many of the arguments made for decentralization point to city or county government as the optimal level of jurisdiction, not state governments (Dahl 1967). In principle, many policies can and have been made at the local level. Prior to the Great Depression, poor relief and welfare were seen to be purely local matters.[1] Until the 1950s, air pollution control in the US was under local, not state, jurisdiction. At the margin, one vote at the state level is as worthless as one vote at the national level. Cities satisfy the representation condition much better. This is particularly true for arguments about the power of citizen participation or deliberation. For most citizens, the state capitol is only marginally less remote than Washington DC.

What criteria can we use to determine whether state, city or federal government is the best level of jurisdiction for making a particular policy? In the public finance literature, transboundary spillovers are decisive in determining the optimal level of jurisdiction for policymaking. Most models assume no negative (or positive) spillovers across borders. Economists William Baumol and Wallace Oates (1988, 295) identified the following conditions under which local standard–setting should be efficient for environmental policy:

1. The effects of the pollution must remain predominantly within the locality (that is, no transboundary effects);
2. The jurisdiction is a price taker for the output it sells and the capital it purchases (that is, no monopoly/oligopoly for output and no possibility of strategic action to attract investment);
3. Policy is determined by the median voter – thus policy is not affected by local rent seeking.

Thus, in the absence of externalities and rent–seeking, under conditions of perfect competition, the standards set by local governments will maximize social welfare. Thus if a locality can regulate pollution which does not spillover, but chooses not to, then this represents the optimal policy choice of that community. In the absence of externalities, that choice should represent an efficient allocation of the community's resources.

On the face of it, the environment is not a natural for decentralization. Most major environmental problems generate damage beyond state borders. Many of the world's pressing environmental problems are global in scope. Traditionally, the economists' ideal has been jurisdictions that correspond to the geographic extent of the environmental externality. However, except for some watersheds, few

---

1   The following were considered to be purely local matters prior at various points in the twentieth century: poor relief, smoke pollution, prohibition, public health, food inspection, sewage treatment and highways.

jurisdictional boundaries correspond with the boundaries of pollution problems. It is unlikely that creating special districts for each type of environmental problem is desirable or feasible (Oates 1972, 49). Most scholars seem content to take existing political jurisdictions as given. The conventional remedy in economic theory was to bump up policymaking to the level of jurisdiction large enough to encompass both the pollution source and the areas impacted by it. The advocates for decentralization argue that when transboundary spillovers occur, they can be adequately addressed without centralized governance, through cooperation and negotiation. This argument will be addressed in Chapter 4.

The Role of State Governments

Current debates on decentralization of environmental policy, as well as other policy areas such as welfare reform, have focused on the state level of American government. The debate takes this form because state governments, unlike county or city governments, have reserved powers that are protected by the US Constitution. With the exception of localities with Home Rule, there are few cities that have powers that could not be taken from them through state legislation. In theory, the states have rights, which are not subject to encroachment or pre–emption by the federal government. However, US Supreme Court decisions and congressional legislation since the New Deal period have permitted substantial federal encroachment (Zimmerman 1991). The Supreme Court under the Bush presidencies has been more sympathetic to states' rights than any court since the New Deal. In the 1990s, the Supreme Court upheld the principle of state sovereignty, ruling that only the federal government, not citizens or companies, can sue a state government for failure to enforce a federal law. This is likely to have profound implications for environmental policy, particularly for citizen suits.

The debate on decentralization concerns the rights of states to make policy autonomously, without federal interference. Because in the US the principle of states' rights was used to justify segregation in the twentieth century (and before that, slavery), it has not been a wholly respectable cause. Thus the case which has been made recently focuses on substantive claims for state level policymaking, rather than simply asserting of the existence of a right. Justice Sandra Day O'Connor's rendition of these claims is as good a boilerplate as any. Federalism, she states, 'assures a decentralized government that will be more sensitive to the diverse needs a heterogeneous society ... increases opportunity for citizen involvement in democratic processes ... [and] makes government more responsive by putting the states in a competition for mobile citizenry'(*Gregory v. Ashcroft* 111 US. S. Ct., 2399–400). She reiterates Judge Brandeis' argument that federalism allows for more innovation and experimentation in government. She also makes the argument that federalism, by virtue of the principle of divided sovereignty, is a check on abuses of government power. If we accept these claims, then the state

Table 2.1 Claims made for noncentralized governance

The claim: Smaller governments/lower levels of government are closer to the people

Type of Argument	The Arguments
Information	smaller means better knowledge of local preferences
	smaller jurisdictions have more accurate information about local conditions
	smaller means easier to monitor
	decentralization leads to greater allocative efficiency because the goods provided reflect local preferences
Preference aggregation	smaller jurisdictions are more homogeneous. the more homogeneous jurisdictions there are, the greater the numbers of people whose preferences are reflected in policy.
	lower levels of government are more democratic
Representation and responsiveness	individual has a greater say – one vote is worth more at lower levels of jurisdiction
	smaller governments are more responsive because they are responsible to fewer voters
	greater public participation/ greater opportunity for participation
	smaller governments are more efficient
Efficiency/ effectiveness	Leviathan hypothesis: smaller governments are more efficient because there are more of them and hence there is more competition
	smaller governments are more efficient because they represent a more optimal size of jurisdiction
	smaller governments are more effective
Innovation and experimentation	state governments are more innovative
	smaller governments are more legitimate
Legitimacy	lower levels of government are more trusted, more highly regarded
	people identify more strongly with lower levels of jurisdiction, hence they feel emotionally closer to these governments

level of government is the optimal level of jurisdiction for policymaking on a wide range of issues.

Claims Made for Decentralization

The arguments made in favour of decentralization are not necessarily grounded in the economists' concept of optimal jurisdiction. What are the arguments in favour of less centralized governance? The arguments are often bundled into a *claim* that lower levels of government are closer to the people. This rather hollow phrase is a useful starting point from which to unpack some of the assumptions about the merits of decentralization. The claims made in favour of policymaking by smaller jurisdictions fall broadly into six types of argument: *arguments about information, preference aggregation, representation, effectiveness, innovation and legitimacy.* Table 2.1 lists the arguments, grouped into these six categories. Many, but not all, of these claims are susceptible to empirical testing.

The Arguments Made in Favour of Noncentralized Governance

'State governments are closer to the people than the federal government.' Taken at face value this is a truism, because unless you live in the Washington DC Beltway, the state capital will always be geographically closer than the national capital. When people say 'government that is closer to the people', what they mean are smaller jurisdictions: smaller in number of inhabitants or in geographic size. It is assumed that governments that are closer to the people will make better policy than governments at a higher level of aggregation. If these smaller governments make better policy, then there can be little justification for allowing higher governments to supersede or overrule these policies or to mandate other policies.

Arguments about Information

The arguments about information are three: lower levels of government have more accurate knowledge of local preferences and the local conditions because of distortions introduced by distance. For the same reason, voters are better able to monitor lower levels of government because of their greater proximity. One of Friedrich Hayek's principal arguments in favour of the market was that it was the best means of discovering and transmitting information: information about preferences as well as supply and demand (1978). For this reason, he favoured smaller jurisdictions: local conditions, local knowledge about those conditions, local preferences. These would combine to permit optimal social choice at the local level, which best reflects the preferences of the residents of that jurisdiction.

These arguments can be tested empirically. On its face, the 'distance equals distortion' argument seems less valid today, because of technology. Technology and techniques such as data mining make possible greater economies of scale in collecting and analyzing information about preferences. Albert Breton, an economist of federalism, argues that:

> ... to defend the traditional line of reasoning, it is therefore essential to be able to argue that more junior governments can acquire information about demand functions at lower costs than can governments located higher up in the system. That may be the case. Casual observation, however, points to economies of scale in polling, canvassing and consulting and to economies of size in interest groups or demand lobbies that convey information on the preferences of their members. Furthermore, cliques, family compacts and other cabals that filter information to governments may be easier to create at more junior levels (1996, 223).

The relative advantage that lower levels of jurisdiction have due to proximity may now be outweighed by greater expertise or resources at higher levels of government, as well as economies of scale.

What about information about local environmental conditions? In theory, lower levels of government should have better information on this, although the caveat about the relative merits of state versus local government would apply. In the case of a problem like odour from large scale hog farms, local government has a much better idea of the extent of the problem than higher levels of government. This is not true of all types of pollution, however. Some types of environmental data, such as dioxin levels, are collected only at great expense. Collecting and analyzing this kind of data requires highly specialized expertise.

In practice, national governments and international organizations are much more likely to collect local data on environmental quality than localities and to disseminate that data. To the extent that data is collected at the local level, smaller governments do not generate extensive, commensurable data, over extended periods of time. This makes comparisons with other localities impossible, which greatly reduces the utility of the data. One example is the data that has been collected under the US *Clean Water Act*. The goal of the *Clean Water Act* was to have US waters fishable by 1983. States and tribal authorities had great discretion in choosing what parameters to measure, where and how often. Because of tremendous discrepancies in how the states carried out this mandate (that is, how many lakes or rivers were sampled), it is not possible to accurately assess the benefits of the billions of dollars authorized for wastewater treatment plant construction in the *Federal Water Pollution Control Act* Amendments of 1972 (US GAO 1986).

The finding above has implications for the next claim: that smaller, more local governments are easier to monitor. If these governments do not generate the data necessary for monitoring by voters, then how can voters monitor?[2] In general,

---

2   One might argue that NGOs can provide data and that government provision is unnecessary. In the case of many data on environmental quality, the costs of data collection

the claim of ease of monitoring depends on how people get their information. If people get their information from face–to–face contact, say at the local barbershop, it is probably true. If voters get their information from news media, then their monitoring capacity depends on available media coverage. Most network newscasts have better coverage of national affairs than state governments. This is particularly true for a specialized beat, like environment reporting.

Arguments about Preference Aggregation

These arguments pertain to what governments do with information about preferences. The claim is that decentralization leads to greater allocative efficiency because the goods provided reflect local preferences. The ideal for revealing and satisfying preferences is the market for private goods. The smaller the jurisdiction, the more closely it hews to the ideal of choice by an individual consumer. However, for public or mixed goods, determining consumers' preferences and willingness to pay is difficult both in theory and in practice. The seemingly small step, which moves from determining the preferences of an individual to determining those of a community, is actually a quantum leap, even for a small community. This move requires aggregating preferences, a process that is redundant under perfect market conditions. Unless the community is known to be homogenous in its preferences, preference aggregation is a complicated matter.

The preference aggregation argument actually hinges on claims about homogeneity, rather than size, per se. It is assumed that smaller jurisdictions will tend to be more homogenous. The more homogeneous jurisdictions there are, the greater the numbers of people whose preferences are reflected in policy. If the entire nation were homogenous in its preferences, there would be welfare losses, not gains, associated with decentralization (Breton 1996, 186). This raises an empirical question, for which we have little data: at what level of jurisdiction is homogeneity of preferences maximized? That is, which level of jurisdiction maximizes the number of people living in homogenous jurisdictions? The answer would depend on the issue. For example, a far greater number of jurisdictions would be required to have homogenous preferences on gay rights versus those on safe drinking water. It is quite possible that on an issue such as standards for safe drinking water, there is little significant variation by region within a nation.

There is little empirical data comparing the policies at the state versus federal level for their correspondence to public preferences. Some studies have tried to measure how well budgets at various levels of government match public demand. The approach taken in these studies has been to estimate public demand for various types of outlays:

---

are prohibitive. Most environmental NGOs do not collect their own data. They primarily analyze data collected by research scientists or government agencies.

> The objective of this approach is to evaluate whether one can account for the
> observed pattern of expenditures by variables from the demand side ... This
> is obviously a very weak test... This approach has been used for the pattern
> of expenditures observed in Germany and Switzerland at various levels
> of government ... The evidence lends some support for the idea that 'true'
> federal states (like the US and Switzerland), which involve a high degree of
> fiscal decentralization, may conduct public affairs more in line with citizens'
> preferences than others (Begg et al. 1993, 50).

This would seem to be fruitful area for future research. The relative
representativeness of national/subnational government policies might vary by
policy type, for example redistributive policies versus those that provide a public
service. This book argues that subnational environmental policies are significantly
less representative of public preferences, even preferences measured at the
subnational level, than national policies. The chapter on Switzerland uses canton
level referendum data to establish this point.

Depending on what the issue is, parsing finely grained distinctions in preferences
may be moot. Economies of scale or the nonexcludability of public goods may
make distinctions meaningless. For example, the preferences of the rather liberal
city of Cambridge MA *vis– à–vis* very blue collar Charlestown MA on the merits
of nuclear deterrence are probably quite different. However, the nature of the
nuclear umbrella means that the sandal wearing pacifists of Cambridge have to
consume (and pay for) the nuclear deterrent, whether they want it or not. It is not
possible to satisfy the preferences of the hawks *and* the no–nukes people, probably
not even at the state level.

The nonexcludability of air quality makes this issue somewhat similar to the
nuclear deterrence example. If we assume significant variations in preferences
for air quality across localities, then many localities will consume excessively or
insufficiently clean air because of the impossibility of supplying the quantity of
the good, which the locality would like.[3]

Arguments about Representation and Democracy

It has been claimed that a system that is more decentralized is more democratic.
William Riker described the debate thus:

> ... [L]ocal government is more responsive to public opinion and more responsible
> to the people' is a typical form of this argument which has been repeated ad
> nauseam in the ideological literature. Fortunately, this particular claim is subject
> to direct investigation. One question is whether or not state governments actually

---

3   In theory, this outcome could be resolved by having people move to locations with
the air quality they like (Tiebout) or by having the downwind people who want clean air pay
the upwind people who prefer dirtier air (Coase). These options are discussed below.

are responsive to democratic control. The recent series of studies initiated by Dawson and Robinson (1963) and brought to a considerable conclusion by Dye (1966) and reviewed by Jacob and Lipsky (1968) generally support the proposition that state governments are more influenced in their actions by the state of their economies than by the demands of their citizens (Riker 1975, 156–7).

More recent scholarship argues that state governments are responsive to public opinion, although the degree of responsiveness varies across states (Erikson et al. 1993). *Statehouse Democracy*, however, does not compare the responsiveness of state legislators with the responsiveness of congressional representatives.

The claim of better representation may, once again, hinge on the homogeneity of the jurisdictions in question, not relative responsiveness of representatives and institutions. Albert Breton (1996, 185–6) writes:

> [T]he problem with the principle of responsiveness is ... the basic intractability of the notion of responsiveness itself. Suppose, for example, that all governments are in some sense equally responsive to the preferences of citizens. Suppose also that the preferences of the citizens are more homogeneous at lower levels in the hierarchy of governmental systems. Then, if governments are equally misinformed about the preferences of their citizens, preferences will not be as well satisfied at higher levels of jurisdiction as at lower ones, even though by construction governments are equally responsive, simply because there is more variability in the distribution of preferences in higher jurisdiction and, as a consequence, more information is needed by senior governance to provide goods and services in volume to generate the same level of utility loss as that generated by junior governments.

An analogous argument may hold for claims of greater responsibility at lower levels. If electoral terms are the same for both levels of government and powers of recall are identical, then it is not clear why a lower level of government is more responsible.

One might test the hypothesis of greater state responsiveness by polling citizens on their preference for their state government *vis–à–vis* the federal government. Recent research suggests that relative responsiveness is not a factor in ranking preferences between one's state government and the federal government. Stephen J. Farnsworth (1999, 85) writes:

> ... support for one's own state government does not spring largely from frustration with the federal government. People generally do not turn to their state capitals because of the sense that Washington doesn't care about them, though economic matters do have some relevance to one's feelings about federalism. One's feelings about one's state government had even less to do with state–specific measures: responsiveness, professionalism, and the partisan or taxation affinity for one's own state. Broad orientations toward the political

world like partisanship and ideology are much more important for predicting
support for the state government than any substantive evaluation of what one's
own state government has or has not done.

Farnsworth also finds that factors predicting low confidence in the federal
government do not necessarily predict a corresponding confidence in state
governments. This is consistent with his finding that these preferences are rooted
in ideology, including a preference for less government at all levels.

Perhaps lower levels of government are more representative because individuals
are better able to influence outcomes. This argument is more persuasive for the town
or city level than the state level of government. The limit at which the marginal
value of one vote approaches zero is a jurisdiction of about 10,000 people (see
data on turnout at New England town meetings below) (Dahl 1967). At the state
or federal level, the marginal value of one vote is negligible. For most states, there
is not much difference between the federal and state levels in this respect: 'the
goal of realizing democratic values to the maximum extent feasible may not be
significantly enhanced by reducing the relevant polity from one of 280,000,000 to
30,000,000 (in California)' (Shapiro 1995, 93).

The greater representativeness of lower–level government cannot be stated as
an iron law. Representation depends on context. In the past, state elections may
have been less representative than Congressional elections due to inequitable
patterns of districting. In many American states, the allocation of seats favoured
rural areas at the expense of urban areas (David and Eisenberg 1961). In some
states, this inequality was enshrined in the state constitution. For example, in 1960
the five largest counties in Florida had half the population and five of 38 (13 per
cent) Senate seats. Los Angeles County had 40 per cent of California's population
and one of 40 seats (2.5 per cent) (Jewell 1989, 86).

Systematic urban under–representation is also evident in Australian states
and Canadian provinces. Legislation, which apportioned seats in Alberta in 1990,
assigned 42 of 83 seats to urban Alberta, even though the urban areas contain 60
per cent of the population. This legislation replaced a 1971 apportionment in which
Edmonton and Calgary, with 52 per cent of the province's population, had only
38 per cent of legislative seats. In many countries, including the US, Canada and
Australia this pro–rural bias is being undermined: '[m]andated malapportionments
are gradually disappearing, often as a result of high court decisions ... or more
generally as the principle of equal representation is given greater importance'(Katz
1998, 252).

Perhaps state and local governments are more representative and more
democratic because they permit greater opportunities for public participation. It is
possible to empirically test the hypothesis that citizens' participation is greater for
lower levels of government. The available evidence indicates that there is no linear
relationship between level of government and participation. In fact, voter turnout
is lower at the state level that at the federal or local level. In North America and
Europe, voter turnout for subnational elections is consistently lower than that for

national elections: 'turnout in mid-term, regional, local and supranational elections – less salient but by no means unimportant elections – tends to be especially poor' (Lijphart 1997). Some scholars argue that low turnout is not a problem: it could mean that candidates already promote the policies favoured by the median voter and thus it is rational for people not to vote (Ledyard 1984).

If small size, personal impact and opportunities for participation are desirable, then clearly the direct democracy of the New England town meeting is the ideal. Joseph Zimmerman (1999) notes that towns in New England and rural Switzerland are the only places in the developed world using the town meeting form of government. However, every year, fewer towns are governed by town meeting, a change due in large part to declining voter interest (Nordell 1999, 12). Attendance at Vermont town meetings has dropped steadily since 1970 (Zimmerman 1999, 95). Across the New England states, attendance is generally higher the smaller the town (Zimmerman 1999, 165). New England towns of under 1000 residents generally have a turnout in the range of 20 to 45 per cent of registered voters. In towns of more than 10,000 people, attendance rarely exceeds 10 per cent and can fall below one per cent.

What are we to make of this? Robert Dahl (1994, 28) has identified this phenomenon as a paradox of democratic representation:

> ... [T]hat larger political systems often possess relatively greater capacity to accomplish tasks beyond the capacity of smaller systems leads sometimes to a paradox. In very small political systems, a citizen may be able to participate extensively in decisions that do not matter much but cannot participate much in decisions that really matter a great deal; whereas very large systems may be able to cope with problems that matter more to a citizen, the opportunities for the citizens to participate in and greatly influence decisions are vastly reduced.

The claim of greater opportunities for participation is, in part, a normative argument. The civic virtue model of government holds that government should promote public and private virtues of the republic, such as participation in public life (Sandel 1995). Yet through a town meeting, New England townspeople have decided to dispose of the town meeting form of government. What does this say about the value of greater individual participation? Are opportunities for individual participation valuable if people indicate that they would rather forgo those opportunities? Should people be compelled to retain the town meeting form of government, even though they do not want to? Would it be democratic to compel people to retain participatory democracy? At some point, this ceases to be an issue about giving people a voice to express their preferences and becomes an issue of designing institutions to shape civic virtue.

Arguments about Efficiency and Effectiveness

Another claim made for decentralized governance is that it is more efficient and more effective. These claims must be differentiated into two groups: claims based on the *number* of jurisdictions and those based on the *size* of jurisdictions. The first type of claim concerns the benefits of competition. The second concerns questions of optimal jurisdiction.

In 1980, Geoffrey Brennan and James M. Buchanan elucidated their Leviathan model of government: governments behave like any other kind of monopoly and act to maximize their surpluses. Brennan and Buchanan's remedy for this problem is competition, which will constrain the taxing power of governments. Thus the greater the number of subunits, the more efficient individual governments will be: '[t]otal government intrusion into the economy should be smaller, *ceteris paribus*, the greater the extent to which taxes and expenditures are decentralized' (Brennan and Buchanan 1980, 185).

This novel thesis does not find strong empirical support, however (Oates 1985). Looking at US county level data, Zax (1988) measured aggregate county public debt and expenditure (for all levels of government in the county) as a share of total county personal income. He found that debt and expenditures were minimized by simultaneously reducing jurisdiction market shares and expanding jurisdiction coverage. Zax concluded that both consolidation and fragmentation had limitations for local government and suggested expanded use of single purpose governments within counties.

The optimal level of jurisdiction varies by type of policy. Here again the absolute size of the jurisdiction may matter more than the level of jurisdiction. Among experts on public finance, there is a consensus that congestible local public goods are best provided by local levels of government.[4] The disagreements arise over exactly what counts as a congestible local public good. For example, is elementary school education a local public good? Although schools are clearly subject to congestion (overcrowding), the quality of the education that children receive has considerable spillovers for the rest of society. In addition, there are economies of scale in goods such as curriculum design.

Wallace Oates' 1972 book *Fiscal Federalism* remains a classic. He sets out two proofs of the Decentralization Theorem: in the absence of 1) cost–savings from centralized provision of a good and 2) interjurisdictional spillovers, the level of welfare will be *at least as high* if Pareto–efficient levels of the good are provided in each jurisdiction than if *any* single uniform level of consumption is maintained across all jurisdictions. If the Pareto–efficient level of the good is the same for all jurisdictions, it does not matter whether the good is provided centrally or not. If nearly all persons have roughly similar demand for the public good and if what

---

4   Congestibility refers to crowding (for example the number of viewers of the same TV program are not subject to crowding, unless they are squeezed into one room trying to watch a single TV set).

diversity exists is spread fairly equally across all jurisdictions, there will be little increase in welfare from the independent local provision of the public good. Note that Oates assumes that central provision implies provision of the uniform level of the good. This is not an inevitable result of central provision, particularly if the central government has information on local preferences and conditions.

On the basis of Oates' purely technical criteria, the arguments for decentralizing environmental policymaking are not particularly strong. First, the environment is the quintessential example of negative spillovers.[5] Second, for goods such as drinking water quality, there is little evidence to suggest substantial variation in public preferences. For goods such as air quality, there may perhaps the greater variability in preferences. However, because of the long–range pollutant transport and the geographical extent of airsheds, there are few cases where air quality can be tailored to local preferences.[6] Third, there are substantial economies of scale in scientific research, which suggests cost savings in centralized environmental standard setting.

In addition to the extent of spillovers and economies of scale in provision, the efficient level of public goods provision depends on the excludability of the good.[7] Economist Rémy Prud'homme (1995, 209) elaborates:

> The standard decentralization model says nothing or next to nothing about production efficiency. The welfare gains to be obtained ... accrue only because supply will better match demand. The hidden assumption here is that supply itself is always efficient. This assumption, derived from the consideration of the private sector (where it does not always hold), is not acceptable for the public sector ... The real issue is whether local provision is more cost–effective than national provision.

Prud'homme ranks 15 local public services on the basis of externalities, excludability and the level of technical knowledge required. He ranks wastewater treatment low on his list of services that can be decentralized to localities. He argues (1995, 216–7) that only highways and railroads are less amenable to decentralization:

> ... street cleaning, water distribution, urban transportation and power distribution appear to be the most interesting candidates [for devolution]. The other end of the spectrum are such services as highways, sanitation, railroads, power production

---

5   Chapter 4 will address the inadequacy of cooperative efforts to manage spillovers.

6   A notable exception would be the Los Angeles basin where most of the air pollution originates in the basin and is retained in the basin, instead of spilling over into other jurisdictions. This is however, a fairly unusual case. Most regions of the United States either export air pollution to or import air pollution from other regions.

7   Excludability refers to the possibility of depriving people of a good unless they pay for it.

and primary education which should be considered for devolution only with
great caution and prudence – if at all.

Arguments about effectiveness and level of government are subject to similar
caveats. There is no unified literature of the effectiveness of service provision for
all types of policy. However it is clear that no simple relationship exists between
centralization/decentralization and effectiveness. The optimal jurisdiction is often
not the lowest possible level of jurisdiction.

In some policy areas, effective reform may require more radical decentralization
than had been anticipated. For example, a World Bank study of Latin American
education reform found education reform was more likely to be successful if
power was decentralized to the level of the school. Decentralization to the level
of the local school board was less effective: '[t]here is growing evidence that at
least some of the characteristics of education decentralization reforms that focus
on school autonomy, as opposed to municipal or regional autonomy, contribute to
higher performing schools'(Burki et al. 1999, 68).

Across the US, both levels of government administer correctional facilities.
John J. DiIulio Jr. has compared the administration of prisons by federal
and state governments. Prison administration is a good basis for comparison
because the goals and responsibilities are roughly equivalent for both levels of
government. DiIulio found significant variation in the quality of management but
this was attributable to differences across specific managers, not across levels of
government. There was no systematic tendency for state institutions to outperform
federal institutions. There was rather more evidence that the Bureau of Prisons
did a better job than its state counterparts. DiIulio argues there is some merit to
centralizing prison administration across the country under the Bureau of Prisons,
which would represent a fairly radical departure from existing practice (Donahue
1997, 128).

Looking at public perceptions of efficiency or effectiveness does not resolve
the issue either. Public opinion polls on perceived effectiveness vary over time and
are not broken down by program area.[8] When in 1972 and 1993, Americans were
asked, 'which level of government gives the most for the money?' the smallest
number of respondents chose state government. In 1972, the greatest number said
the federal government gave the most for the money. In 1993, the largest number
of respondents said local government gave the most for the money (US ACIR
1993).

It is worth keeping in mind that the size, shape and number of states or
provinces in a nation are the product of historical accident. If Rhode Island did
not exist, would we need to invent it now? There is nothing 'natural' about most
of these jurisdictions. If jurisdictional boundaries coincide with meaningful

---

8   We cannot say anything about this trend in the United States between 1976 and
1996 because these questions were excluded from the American National Election Studies
during this period.

boundaries in the natural world, such as biotopes or watersheds, this outcome is usually a coincidence. Writing in the 1930s, legal scholar Karl Llewellyn (1934, 38) mused: 'perhaps this ... does not warrant decentralized authority along the lines drawn by state boundaries.' Llewellyn conceded that, although somewhat arbitrary, some kind of line was necessary for dividing powers between federal and local governments. However, he did not think it self–evident that existing state boundaries afforded 'a sound basis for revising an "inane" pattern of distribution of functions' (Shapiro 1995, 43).

There has been no systematic study comparing the effectiveness of federal vs. state and local provision of environmental regulation. One reason is that very little comparative data is available on the performance of state and local governments in this area, with the exception of programs that have been delegated by the US federal government. There has been a comparative study of natural resource management, in American forestry. Tomas Koonz compared how federal and state forests were managed (2002). He found that state forests are managed for timber yields and revenues, whereas federal forests are managed for conservation. Furthermore, contrary to claims about state governments being 'closer to the people', federal managers were more likely to encourage citizen participation. The differences between the two levels could not be explained by differences in the beliefs of bureaucrats, which were fairly similar across the two levels. His findings were consistent with the hypothesis that lower levels of government are more vulnerable to economic pressure.

The theoretical arguments on public goods provision do not present a strong case for decentralizing environmental policy. Air and water quality are generally neither excludable nor congestible goods. Increasing the number of people in a city does not reduce the amount of air available to breathe. The number of factories spewing smoke into the air does not limit the capacity of additional factories to dump their waste into the air, although obviously it has consequences for other uses of that air. The provision of public water supplies is excludable. However it has proved to be very difficult to exclude people from dumping waste into water. Lastly, there are tremendous economies of scale in scientific research and environmental standard setting requires substantial scientific evidence and expertise.

The Innovation Argument

Throughout his career, United States Supreme Court Justice Brandeis argued that federalism was beneficial because it creates a multiplicity of natural laboratories, which allow for local innovations to be studied and possibly copied by other jurisdictions in designing their own policies.[9] Decentralization can permit innovation. Because there are many more points of decision–making, there will be greater variety of approaches. Ideally, these approaches represent optimal solutions for that jurisdiction.

---

9   Dissenting opinion in *New State Ice Co. v. Liebmann* 285 US 311 (1932).

For individual jurisdictions to serve as natural laboratories, the benefits and costs ought to be internalized to the jurisdiction. If the programs being provided, such as education, are not local public goods, we should expect underinvestment and free riding. In the case of policymaking in a single jurisdiction, one may draw false conclusions through the fallacy of composition. For example, one state could slash its welfare rolls by giving clients one–way Greyhound bus tickets out of the state. However, it is not possible for *all* states to slash their welfare rolls by emulating this 'begger they neighbour' practice.

In environmental regulation, there is little evidence that subnational governments with more autonomy will be more innovative. On the contrary, the evidence points to the conclusion that less centralized systems are *less* innovative and there is no inevitable diffusion of those innovations that do occur. There are few incentives for being the first to regulate. Even when large states act unilaterally, there is no guarantee that other states will follow. Barry G. Rabe's work (1995, 1998) comparing the United States and Canada finds little support for the hypothesis that greater decentralization leads to greater innovation. Rabe found far more innovation and awareness of innovations in other jurisdictions in the relatively centralized US than in Canada, which is highly decentralized.

While decentralization may, in theory, permit innovation, it does not necessarily promote it. The empirical literature comparing Canada and the US on environmental policy innovation is presented in the Canadian chapter. Environmental policy in the US, subject to minimum federal standards, is far more innovative than provincial environmental policies. In addition, state environmental agencies show more evidence of learning from others than the Canadian provinces do, even though the provincial governments have far more discretion in making environmental policies than the state governments do. Finally, even if decentralization did lead to innovation, is maximizing innovation intrinsically good? Suspending all existing environmental regulation, and not replacing it with anything, would be highly innovative. We cannot assume that policy innovation is identical with technological progress or better environmental policy.

Arguments about Legitimacy

The final component in the 'closer to the people' argument is legitimacy. It is argued that lower levels of government are more legitimate. This greater legitimacy derives from assumptions about a higher level of trust and/or a closer emotional identification. Some elements of this argument are susceptible to empirical testing. Other claims of legitimacy, however, are essentially pure value judgements and cannot be assessed on the basic facts.

On the matter of trust in government, poll data generates conflicting results, particularly over time.[10] In 1995, a majority of the Americans polled reported

---

10   We cannot say anything about this trend between 1976 and 1996 because these questions were excluded from the American National Election Studies during this period.

trusting their state government more than the federal government. Most groups surveyed, including Democrats and Republicans, liberals and conservatives held this view. Conservatives were much more likely to trust their state government more than the federal government. In general though, Americans had little trust for any level of government:

> the advantage held by state governments in comparison with the federal government should not, however, be read as a strong endorsement of state governments. Only about one in three Americans say that they trust their own state government to do the right thing always or most of the time (Blendon et al. 1997, 209).

Factors predicting low confidence in the federal government do not necessarily predict a corresponding confidence in state governments (Farnsworth 1999).

Of particular relevance to this book, is the finding that despite the greater trust in state government, the majority of Americans polled still wanted to the federal government to have greater powers than the states for certain policy areas:

> ... although the majority of Americans trust their own state government more, the public still prefers that the federal government have more responsibility than state governments for certain problem areas ... [A]bout two–thirds think that the federal government should have more responsibility than state governments for strengthening the economy and protecting civil rights. More Americans prefer that the federal government, rather than state governments, have responsibility for protecting the environment (50 percent for federal, 38 percent for states) ... (Blendon et al 1997, 212).

Even when the American public trusts one level of government more than another, they are not prepared to write a blank cheque to the more trusted government. The finding of a preference for federal, over state, environmental policymaking is also reflected in Swiss referendum results over many decades. Canadian polls on this subject, however yield ambiguous results, suggesting distrust of both federal and provincial levels of government. However, Eurobarometer polls over several years show that the majority of citizens in European Union member states prefer environmental policy to be made at the level of the EU, rather than the member states.

In general, arguments about legitimacy in assigning jurisdiction for policy areas are somewhat specious. They are generally employed as a post hoc rationalization of an existing division of powers. Powers or government functions are not reallocated on the basis of which government is most trusted or most popular at a given point in time. In most federations, the division of powers is determined by the Constitution, subject to judicial interpretation. Even if the public were to have clearly defined preferences with regard to the allocation of powers, it is difficult to fulfill these preferences through the political process. Switzerland would be the notable exception to this rule. In Switzerland, most national referenda serve to reassign powers within the federation. In a federal system, allocations of powers are typically decided in court, or sometimes at the end of a gun, as they were during the American Civil War.

Arguments about legitimacy or greater emotional attachment to a particular level of government are often arguments about sovereignty. Sovereignty is defined as the belief that a particular jurisdiction has an inalienable right to make laws without interference from other governments, without being compelled to act by another government. Claims of sovereignty are not falsifiable and cannot be resolved by resort to empirical evidence. For example the nineteenth century Southern secessionist John C. Calhoun believed the states were inviolate organic entities and the US was nothing more than a voluntary compact between the states. This is fundamentally a normative position. If sovereignty originates in a popular mandate, then it is not clear that one level of elected government has a greater claim on sovereignty and legitimacy than another.

Finally, these claims of sovereignty through legitimacy privilege geography over all other identifiers. In the eighteenth and nineteenth centuries, it may have been reasonable to define community purely geographically. In the age of the Internet, this definition does not stand up to scrutiny. Individuals have always had multiple attachments (and identities and loyalties) on the basis of sex, religion, language as well as geographic residence. However, it is no longer the case that the strongest attachment is geographically concentrated. For example, homosexuals living in Mississippi may feel a greater identification with the national or international gay community than with their neighbours and other Mississipians.

Conclusion

The existence of externalities violates one of the basic assumptions in favour of decentralization. However, advocates of decentralization make claims on many dimensions. The claims made for decentralization with regard to better information or efficiency do not find very strong support in the empirical literature. The most persuasive claim is that lower levels of government better represent preferences. However, lower levels of government may be incapable of providing a particular good due to the nature of the good (for example, New Jersey providing good air quality without action by New York). More importantly, if environmental policymaking at the subnational level is characterized by a Prisoner's Dilemma, citizens have less chance of seeing their preferences enacted into policy locally than nationally.

Chapter 3

Interjurisdictional Regulatory Competition and Fears about Competitiveness

Economist Albert Breton, a leading scholar of the economics of federal arrangements, has argued that competition among governments in federations is beneficial, in the same way that competition between firms in the private market maximizes economic welfare (2002, 36). However, Breton has also identified 'races to the bottom' or competition in laxity as one case of decentralization failure, where decentralization of a policy does not produce a socially optimal outcome.[1] When is regulatory competition between governments harmful? How often is it harmful?

The issue of harmful competition is particularly important in environmental regulation. It appears governments are hypersensitive to the possibility of impacts on competitiveness whenever environmental regulations are contemplated. This is so, even though economists have found little empirical support for these concerns. These concerns, though baseless, appear to form a substantial impediment to pollution control and the impediment is especially pronounced at lower levels of jurisdiction.

Concern about the competitive impacts of environmental regulation is as old as the Industrial Revolution. When the British Parliament debated the first smoke control bills in the 1840s, industrial interests argued that smoke control would cause unemployment in northern industrial towns (Ashby and Anderson 1981, 13). The failure of British local governments to enforce smoke control was justified by reference to growing economic competition from Germany and America (Flick 1980, 37–8). Fears of competition in laxity between jurisdictions also arose early.

Over many decades, the British government appointed several commissions of inquiry on both air and water pollution. Most of these commissions recommended uniform national laws and a national inspectorate to enforce them, 'removed as far as possible from all local influence', which was the recommendation of a House of Lords Select Committee on Injury from Noxious Vapours (Dingle 1982). However, these recommendations were not heeded, except for some toxic industrial air pollution. Writing on river pollution in nineteenth-century Britain, one author observes:

> ... a pervasive fear of the cost and economic consequences of anti-pollution policies and practices. This acted as a bar to effective action even where harm

---

1   Breton prefers the term 'dynamic instability' to 'race to the bottom'.

from river pollution was acknowledged. Local authorities, for example, held back from the adoption of anti-pollution schemes for sewage disposal partly because they did not want to disturb the rate payer. The rate payer then, as in the case of the taxpayer now, could be easily aroused by the threat of increased rates. Manufacturers and mine owners feared that a River Pollution Prevention Act would mean prohibitory costs, loss of profit, and give the advantage to competition elsewhere in Britain and, in the case of some manufactures, foreign quarters. Decline of trade and loss of jobs were twin cries raised at the least sign of interest by Parliament in acting to prevent river pollution. Governments remained ever sensitive to Britain's dependence upon manufacturing and trade, and this both delayed passage of anti-pollution legislation and, when enacted, resulted in a flawed and weakened measure (Breeze 1993, 42).

Industrial interests in the UK lobbied very strongly against central regulation and enforcement, in favour of regulation by local authorities. This opposition was successful until the second half of the twentieth century.

Reading accounts of the nineteenth century struggle with pollution, it is striking how little the issues have changed. The same concerns recur: if Mexico has no environmental standards, then the US must lower its standards or Mexico must adopt US standards. There are the same battles over local versus central regulation and enforcement. Local politicians argue in favour of deregulation and limited environmental standards to protect local jobs. Those who favour stringent measures argue that local control will lead to competition in laxity, or at the very least, laxity in enforcement of whatever regulations are in place.

The persistence of these debates in our time is puzzling given the lack of evidence to support claims that higher standards of environmental protection lead to lost exports and jobs or factory closures. Where the effects of environmental regulation have been quantified, the impact on jobs is very small in comparison with wage differentials or interest rates, for example. There is no evidence for a robust tradeoff relationship between growth and environmental protection, particularly not for the economy as a whole (Goodstein 1999). For new businesses, labour force cost and quality, distance to major markets and infrastructure are the most significant factors in business location decisions. And yet, governments still strive to limit environmental policies and push for less stringent measures.

This chapter presents the relevant literatures on interjurisdictional competition. The first section presents the theory that relates to the questions: Do governments compete? If yes, is this competition harmful? The section sets out the theoretical models of interjurisdictional competition, in which governments compete for residents or for investment. The second section presents empirical evidence on cases of interjurisdictional competition and assesses the role of externalities in determining harm or benefit. The third section presents the empirical data that finds no reduction in competitiveness as a result of environmental protection. The fourth section sets out the puzzle of legislator and regulator behavior, in light of the data showing no economic harm from more stringent environmental policies. The final section puts forward some hypotheses, which might account for the discrepancy between what economists know and how legislators appear to behave.

Competition and the Case for Decentralization

Some arguments in favour of decentralization depend on competition. For example, Brennan and Buchanan's Leviathan hypothesis (1980) assumes that a greater number of jurisdictions is a brake on taxation, because there is more competition between the jurisdictions. Competition is implicit in variations of some of the other arguments. US Supreme Court Justice Louis Brandeis argued that decentralization created more possibilities for innovation because of the greater number of jurisdictions. His argument is often recast as one where competition between the larger number of jurisdictions compels more innovation at lower levels of government.

John Shannon (1991) combined the Leviathan hypothesis with a competition–enhanced Brandeis argument to create a convoy analogy of American federalism. Competition for investment constrains the states' power to tax. At the same time, competition for residents means that states cannot fall too far behind the pacesetter states in providing new public services. There is, however, considerable debate on whether governments actually do compete, and if so, what the effects of competition are.

In markets, competition between firms to sell private goods to consumers generates an outcome that maximizes social welfare. Under conditions of perfect competition, this outcome cannot be improved upon. If governments compete, is social welfare maximized? There are two major economic models of interjurisdictional competition. According to the Tiebout model, governments compete to attract mobile citizen–voters. Under the fairly restrictive assumptions of the Tiebout model, this results in a welfare maximizing provision of public goods. According to Richard Revesz's model (1992, 1211–2), jurisdictions with immobile residents will compete to attract mobile firms, which he argues maximizes welfare. When the assumptions of these models are relaxed to make them more representative of the real economy, competition between jurisdictions does not necessarily maximize society's welfare.

The Tiebout Model

Charles Tiebout (1956, 419) conceived of his model as a rejoinder to Paul Samuelson's model of pure public goods provision. Samuelson argued that there is no voluntary way to get people to honestly reveal their willingness to pay for a public good, because of the incentive to freeride once the public good has been provided. Therefore public goods are a type of market failure and must be paid for by (involuntary) taxation. Tiebout's insight was that revealing preferences for public goods was not a problem if the goods were *local* public goods: people would vote with their feet, moving to jurisdictions that provided them with the local public goods they want. In addition, citizens were assumed to be perfectly informed, perfectly mobile and to have a large number of jurisdictions to choose from.

Tiebout's idea has subsequently been extended as a generic argument in favour interjurisdictional competition. A Pareto optimal outcome is attained through citizen mobility and the competition of jurisdictions for citizens. There are those who extend Tiebout's argument from the local government level to state governments. James M. Buchanan (1995) argues that states are forced to compete with one another in providing public services at the lowest possible cost because at least some citizens retain the exit option. Thomas Dye (1990) argues that state governments are better able to satisfy preferences, than the federal government, by competing to attract residents.

When Tiebout's name is bandied about, it is often assumed that jurisdictions compete to attract residents the same way that private firms compete for customers. John D. Donahue (1997, 121) laments that:

> ... the theme of marketlike efficiency through intergovernmental competition can be telegraphed, among the cognoscenti, by merely mentioning Tiebout's name. A teeming chorus echoes the theme, though often with most of the caveats and conditions stripped away.

Despite the fast and loose way that Tiebout is invoked, a Tiebout equilibrium is only equivalent to a market for private goods if there is a large number of jurisdictions. In the market for private goods, firms will innovate and lower prices in response to competitive forces. In Tiebout's model, however, jurisdictions that have attained optimal size will not seek to attract additional residents.

It seems unlikely that the Tiebout model applies to environmental problems, other than extremely local ones such as noise pollution. Tiebout assumes there are no positive or negative externalities. Clearly this last assumption does not hold for most aspects of the environment. There is ample evidence of a correlation between property values and environmental quality, particularly within urban areas, but this pattern of capitalization into land values in no way proves the existence of a Tiebout equilibrium or an efficient level of public goods (Ridker and Henning 1967).

The Revesz Model

Richard Revesz (1992, 1211–2) argues that:

> ... competition among states for industry should not be expected to lead to a race that decreases social welfare ... such competition can be expected to produce an efficient allocation of industrial activity among the states.

His argument is broadly consistent with the model of interjurisdictional environmental competition set out by Wallace Oates and Robert Schwab (1988).

Their model, which finds interjurisdictional competition to be efficient, has the following assumptions:

- several jurisdictions are competing to attract a fixed amount of capital;
- pollution is purely local, with no transboundary externalities;
- residents are homogenous and immobile within the region;
- costs and benefits are internalized to the jurisdiction because all citizens work in the polluting industry;
- the industry has constant returns to scale;
- there are no profits;
- governments are welfare maximizers;
- policy is determined by the median voter.

Under these conditions, local governments make an optimal trade–off between pollution and the benefits of attracting investment. If some of Oates and Schwab's assumptions are relaxed, however, the outcome is no longer welfare maximizing. If pollution spills over into other jurisdictions, the outcome will not be efficient. If the governments are assumed to be budget maximizers, rather than welfare maximizers, they will tend to set excessively lax environmental standards.

Economist Eric Levinson (1997) has contrasted the findings of the Oates and Schwab model with the results obtained by Markusen, Morey and Olewiler. In the Markusen, Morey and Olewiler model (1995), competition between two regions for a single firm with increasing returns to scale results in suboptimally lax regulation. In this model, the optimal regulation is that set for both regions by a central government. Levinson concludes that the efficiency of outcomes depends on the pattern of ownership, that is, where do the owners reside? If locally owned firms compete in efficient capital markets to attract financing, then local jurisdictions will regulate efficiently. However, if jurisdictions compete to attract enterprises, which are owned outside the jurisdiction, they faced distorted incentives to tax and regulate. This distortion leads to suboptimal levels of environmental quality. In addition, Levinson notes that several of assumptions in these models are unrealistic: no pollution spillovers, welfare maximizing governments and homogeneous citizens in each region.

Empirical Evidence of Regulatory Competition

What is the real world evidence on regulatory competition? Does regulatory competition occur, and how often? The issue has often been framed as a 'race to the bottom' in standards. Races to the bottom are a subcategory of regulatory competition. While a race to the bottom is an example of regulatory competition, not all regulatory competition results in a race to the bottom. The absence of a race to the bottom does not mean that optimal policies are in place. The more appropriate question is how policymakers are constrained by the actions, or anticipated reactions, of other policymakers.

A race to the bottom implies regulatory competition but the opposite is not true. Not all cases of regulatory competition manifest as races to the bottom. One reason why a race to the bottom should not be taken as the ultimate test of competitive pressures is that, if competition is extremely intense, we would not observe a race to the bottom because environmental taxes or regulations would *never get off* the bottom. 'Stuck at the bottom' would be a better description. The bottom (a low tax or regulation) would be the equilibrium outcome. A *race* would only be observed if a different equilibrium existed and there were some exogenous shock which led the system to find a new equilibrium point. For example, no race to the bottom has been observed in carbon taxes because there are significant disincentives to the introduction of carbon taxes.

Most scholars consider races to the bottom to be rare, although Albert Breton (2002, 39) argues that, if races such as currency devaluations and tariff wars are included, they are not infrequent. There are documented cases of races to the bottom in regulation, although the harmfulness of these races is debated. The race ensues when the prior equilibrium level of regulation is subjected to a shock.

William Cary (1974) used the term 'race to the bottom' to describe how American states reduced shareholder protection in order to encourage more corporations to incorporate in their state. In the case of interstate competition over corporate chartering, the trigger was Delaware loosening its standards, in order to earn more revenues for the state government. Other states reduced their levels of shareholder protection in order to attract corporate charters. A very similar competition for corporate charters occurred in the nineteenth century, triggered by New Jersey (Shugart and Tollison 1989).

At the international level, safety regulations in shipping have seen competition in laxity. Dale Murphy (2004) described how, in the case of shipping standards, the shock was the creation of flags of convenience. Less developed countries, such as Liberia, specialized in registering ships in order to gain revenues for the government. In return, these countries imposed very few criteria for seaworthiness or worker safety or environmental issues. Murphy shows how as the share of world tonnage registered under flags of convenience increased, so did the total tonnage lost in shipwrecks. In response to the defection of national fleets to flags of convenience, some OECD countries (such as Norway) introduced parallel shipping registries with less stringent standards.

Are Externalities the Only Problem in Interjurisdictional Competition?

In the case of corporate charters and flags of convenience, competition has clearly taken place. But not everyone would agree that the outcome of this competition has been harmful. Arguments about harm from regulatory competition usually center on the existence of externalities: costs imposed on parties outside the transaction. In the case of state incorporations, Ralph Winter Jr. (1977) argued that the competition is efficient. He claimed that those owning shares from firms incorporated in Delaware have already factored in the level of shareholder protection into the price they paid for their shares. Thus shareholders are not subject to an externality. A similar argument could be made for flags of convenience in shipping. The ships' owners, the owners of cargo and ship's crew all know the ship under a flag of convenience is more likely to sink than a ship under a regular registry. They can all factor this risk into the price they pay or the wages they receive.

In theory, interjurisdictional regulatory competition is not a problem, provided there are no transboundary externalities. According to this view, differences in regulations are simply price differences across jurisdictions, similar to differences in wage rates. Mitigating transboundary externalities, such as pollution, is considered to be legitimate and economically beneficial. Regulatory differences without transboundary externalities are pecuniary externalities and attempting to address this is tantamount to price–fixing. It appears however, that even for regulations without transboundary considerations, governments are very concerned about regulatory differences and the price difference implicit in this.

It should be possible to set regulations locally, however, it may be very difficult for local governments to act, because of concerns about competitiveness. In this way, policies made at the local level may not reflect local preferences. The citizens of a country may have to live with suboptimal policy outcomes because each individual subnational jurisdiction is unwilling to act alone, because of competitiveness concerns. For the local governments to act, they need assurances that their competitive position will not suffer as a result. For this reason, they may prefer a level regulatory playing field, set by a higher level of jurisdiction. Economists generally reject demands for regulatory harmonization, particularly if no externalities are involved (see Leebron 1996).

The history of child labour in the US in the early twentieth century suggests that competition can lead to undesirable results in the absence of externalities. In theory, employing children in factories imposes costs primarily within the state, rather than on other states. The cost of children being maimed in mining accidents or growing up stunted and illiterate, is borne by the children themselves, their families and the jurisdiction they live in. However, in the US, it was almost impossible to regulate child labour while the issue remained under state government jurisdiction. Even though opinion polls from the time indicate that a majority of Americans favoured action on child labour, the states were unable to act individually or in cooperation. Sixty–one and 76 per cent of Americans polled in 1936 and 1937, respectively, favoured a Constitutional amendment authorizing the US Congress to regulate and prohibit the labour of those under 18 (Gallup Institute 1972, 23).

Some northern states went so far as to impose a tariff on products from states using child labour, a measure which was struck down as a violation of the Interstate Commerce Clause. Interstate compacts on child labour and minimum wages, negotiated during the Depression, were dismal failures. The states that made the greatest use of child labour did not join the compacts. Even the few north–eastern states that joined the compacts found their legislatures unwilling to ratify them. From an economists' point of view, it is illogical that states would not restrict child labour in their own state because other states, such as North Carolina, refused to restrict the practice. Yet the states were clearly motivated by concerns over competitiveness.

Similar forms of competition can be seen in environmental issues. Jurisdictions will compete on environmental regulations where the form of damage is purely local. For example, the German Länder and municipalities fought maximum national standards on noise in residential areas, out of fear that it would disadvantage the local economy (Posse 1986, 83–4). It is not clear why a maximum national noise standard would disadvantage local economies, if everyone had to adhere to the same ceiling for noise pollution.

There are few types of pollution which are purely local, other than noise. Sewage treatment is more local than most environmental issues, particularly for cities on a lake or very large cities on a river. Assuming that these jurisdictions obtain their drinking water from the same body of water that they dump their sewage into, they stand to capture substantial gains from any investment in treating their own sewage. However, even under these conditions, localities may be extremely reluctant to invest in treatment. Local governments are concerned about the impact on the local tax rates, which in turn might affect the locality's attractiveness to potential investors.

The US chapter describes the failure to clean up Lake Tahoe, which lies between California and Nevada, despite the threat which sewage pollution posed to the area's largest industry, tourism. Fritz Scharpf describes a similar unwillingness by communities in the German *Land* of Baden–Wurttemburg to invest in municipal sewage treatment to prevent eutrophication of Lake Constance, which is the major source of drinking water for the *Land* (Hanf and Scarpf 1978, 79). Even when cities or towns stand to capture most or all the benefits from environmental protection, there is still great reluctance to act on these matters.

Harmful Interstate Competition to Attract Investment

Interjurisdictional competition to attract new investments is, in many respects, analogous to competition over environmental regulations. In theory, state and local governments should offer incentives based on a careful calculation of tradeoffs. A survey of the literature found:

> ... neither the literature reviews nor the empirical studies included within this volume support the notion that state and local governments can effectively alter the geographic distribution of industry through their various fiscal offerings and inducements ... there appears to be near-unanimous agreement among the contributors that industrial location decisions are dominated by the following factors: markets, labour quality and transportation accessibility (Moore et al. 1991, 269).

Despite this evidence that incentives are ineffective in influencing location decisions, governments continue to offer them. As with competition on environmental regulations, a net loss in social welfare is not sufficient to prevent competition.

Many politicians realize that they have no control over the variables that most strongly affect business location decisions, such as distance to major markets. Even where politicians have some control, such as over the quality of education, these efforts take decades to bear fruit in terms of better jobs and more of them (as in North Carolina). Studies have shown that these incentives are not effective in changing industry location decisions and simply represent a net wealth transfer from tax payers to firms. Bidding is a zero sum game or even a negative sum game, as governments systematically overbid to attract firms (Buchholz 1998). As a result, these incentives represent a simple income transfer from taxpayers to private firms.

The harmful outcome results because of information asymmetry. The firms know their true preferences with regard to locations but governments do not. The firms are able to play one jurisdiction off against another, thereby extracting substantial tax incentives. States are unable to stop themselves from participating in this competition. Scholars studying this issue have called for national policies to end the Prisoner's Dilemma (see Buchholz).

Competition and the Concern About Competitiveness

There is a puzzle here. Economic theory indicates that capital mobility and trade should not lead to suboptimal environmental standards. Economic studies find that, even in highly pollution intensive sectors, pollution control costs rarely exceed ten per cent of total costs. The econometric evidence indicates that jurisdictions with lower environmental standards do not gain an edge over others in attracting new investment. Surveys of executives find that environmental regulations are quite low on rankings of factors affecting location decisions.

Yet this book finds that sub–national jurisdictions systematically tend to have fewer and more lax environmental standards than national governments do. Anecdotal evidence indicates that governments are very concerned that environmental protection will harm the competitiveness of their economies. The paradox is that governments appear to compete on environmental regulations and yet gain no economic or employment benefit from doing so. The net result for Canadian provinces or American states is the same number of firms distributed in roughly same way, but with more total pollution. This represents a loss of social welfare. This outcome sounds highly irrational, and it is. It results from a collective action problem wherein all jurisdictions would be better off if they cooperated, but they do not and as a result, all are worse off.

**Table 3.1 Hypothesized impacts of decreased competitiveness attributed
to environmental regulations**

	Domestic effects	International effects
Declining production	reduced output	increasing imports and declining exports of pollution intensive goods
Capital flight to less stringent jurisdictions	plant relocation, plant closures, job losses	increased Foreign Direct Investment (FDI) in pollution havens
New investment forgone	lower rates of new plant 'birth' in more stringent jurisdictions	increased FDI

This section aims to persuade the reader that there is no evidence for economic gains to be made through weak environmental policy. The section surveys the existing literatures on the hypotheses set out in Table 3.1 above. The first is the literature on environmental regulations and competitiveness or industry productivity. The second segment addresses findings on environmental regulation as the cause of plant closure and job loss. Third is the literature on domestic industry location decisions: surveys, econometric studies and material for site selection professionals. Last is the literature on the relationship between environmental regulation and trade and, more narrowly, pollution havens abroad.

When Ronald Reagan became President, the cost of regulations to the American economy began to receive a great deal of scrutiny. Environmental regulations were no exception. The 1970s had seen a serious slump in the US economy, particularly in the rate of productivity growth. Murray Weidenbaum, Chairman of Reagan's Council of Economic Advisers, is a representative critic of the federal regulatory burden. He argued that regulations reduce corporate innovation, reduce new capital investment, contribute to unemployment and undermine the entrepreneurial nature of the capitalist system. He estimated the annual cost of energy and environment regulations at \$8.4 billion, in 1976 dollars (Weidenbaum 1979). He did not attempt to estimate the benefits to society of improved environmental quality. He claimed that in the early 1990s '[e]nvironmental regulations cost each family more than \$1,000 a year'(1992, 40). In articles such as 'A Neglected Aspect of the Global Economy: the International Handicap of Domestic Regulation', Weidenbaum (1995) continued to argue that American industry is hobbled by regulations and that this threatens US exports.

Politics makes strange bedfellows and the politics of environmental regulation is no exception. It must be one of the few issues where radical environmentalists and conservatives are in agreement: domestic US environmental regulations place US firms at a disadvantage in the global marketplace. The left and the right part company with regard to solutions: those on the left favour protectionism or tariffs, those on the right would like to see deregulation in the US. One of the peculiarities of this left/right consensus of opinion is that there is very little empirical support for it.

The myth, however, refuses to die. Arik Levinson, who has conducted econometric studies on the issue, concluded that:

> [w]hatever the reason, there remains a large gap between the popular perception that environmental regulations harm competitiveness and the lack of economic evidence to support that perception. I suspect that the existing literature cannot convince policy makers or the public that links between environmental regulations and industrial location are insignificant (1996, 453).

The studies find no evidence that environmental regulations significantly affected productivity growth, neither for the US economy nor for other OECD economies surveyed (US CBO 1985). The costs of pollution abatement were found to be significant only for a small subset of industries, such as plastics, primary metals, pulp and paper, petroleum and utilities. Barbera and McConnell (1990, 50) found that, for five of the most heavily polluting US manufacturing industries (paper; chemicals; stone, clay and glass; iron and steel; nonferrous metals) environmental regulations accounted for 10–30 per cent of the productivity decline of the 1970s.

The Myth of a Tradeoff Between Jobs and the Environment

Many estimates of regulatory costs and of job losses caused by environmental protection are poorly substantiated and politically motivated.[2] Economist Eban Goodstein (1999, 66) has reviewed the studies on environmental regulations and employment in the US and concluded that:

> [l]ocal job losses due to environmentally related plant shutdowns are simply tiny compared to the real downsizers: technology, trade and corporate restructuring. Through the magic of public relations, layoffs on the order of 1,500 per year, spread out over half a dozen plants nationwide, have transformed environmental regulation into a giant job killer.

Goodstein analyzed data on mass layoffs between 1995 and 1997 and found that layoffs which occurred because of environment or safety–related factors never amounted to more than 0.3 per cent of annual layoffs in the US. Averaged over the

---

2 For example, despite the provocative title, *Measuring the Employment Effects of Regulation: Where Did the Jobs Go?* contains no data on projected job losses due to environmental regulation and very little on other types of regulation. The focus of the book is the federal government's administrative procedures and its failure to implement President Reagan's policies in this area (Zank 1996).

three years, the 'environment related' reason for layoffs was the smallest category of the 23 reasons surveyed. Goodstein notes that conflict between environment and jobs is starkest in extractive industries, such as logging and coal mining, not in manufacturing.

The belief persists that environmental regulations contribute to unemployment and force factories to close, putting people out of work. There is no empirical support for the position that environmental protection negatively impacts jobs. An OECD report on employment and environment (1997, 100) concluded that:

> ... environment–related job losses during the past two decades look almost irrelevant in comparison with job losses resulting from other corporate decisions and government policies (e.g. automation of plants, foreign investment, budget cuts, or from substantial changes in exchange rates). Finally, jobs are more likely to be at risk where environmental standards are low and no innovation in terms of cleaner technologies is taking place. Overall, the net employment impact of environmental policies likely to be positive but small.

The OECD report drew on findings from European Union countries (particularly Germany), and Canada as well as the US.

Nor are environmental regulations the decisive factor in the shutdown of older plants. In most jurisdictions, environmental regulations benefit existing plants at the expense of new facilities (Buchanan and Tullock 1975). In most jurisdictions, environmental regulations are grandfathered: existing facilities have to meet less demanding standards than facilities which have yet to be built. Thus, such grandfathered regulations constitute a barrier to entry in the sector.

Furthermore, there is evidence that plants at risk of being shut down are subject to more lax regulatory enforcement. For example, steel plants in danger of closure were subject to less rigorous enforcement by the EPA than new facilities, as were plants which were major employers in the community (Deily and Gray 1991). Matthew Kahn (1997) found that facilities located in *Clean Air Act* nonattainment areas were *less* likely to close down than those in *Clean Air Act* attainment areas.[3] Robert Crandall (1993) found that, while environmental regulation seemed to reduce job growth in manufacturing plants, it was also associated with fewer layoffs during economic downturns.

The Perceptions of Business People

Surveys have asked business leaders which factors affect their location decisions or have asked executives to rank a series of factors. Executives mention environmental regulations but appear to be overestimate the costs regulations impose. In an

---

3   *Clean Air Act* nonattainment areas have yet to meet federal standards of local air quality whereas attainment areas already meet the minimum standard for air quality.

assessment of these survey studies, Robert Tannenwald (1997) found that, at most, environmental policies had a moderate impact on business leaders' location decisions.

Southern California's economic woes in the early 1990s are a good example of the disjuncture between perceptions about regulation and employment or relocation, and empirically grounded facts. The Southern California economy lost 600,000 jobs between July 1990 and December 1993. The business press and polls of California business people identified California's clean air regulations as a major cause of these difficulties (Kerwin and Grover 1991, 44–45). In 1991, the California Business Roundtable asserted that, because of California's numerous business taxes and environmental regulations, one quarter of the state's manufacturers were planning to relocate ('Raising Business Costs' 1991).

A study of air quality regulation and economic performance in California found that California's business climate could not substantially account for the job losses. Rather, job losses resulted from industry specific factors (such as declines in the aerospace and defence sectors) and broader economic trends, not business relocations (Hall et al. 1997). California's taxes, laws and regulations are fairly similar throughout the state, yet 80 per cent of the job losses were in the Los Angeles area. Sectors which had low air pollution compliance costs show substantial job losses. In contrast, those sectors with high compliance costs (utilities, petroleum products, oil and gas, auto repairs), which account for 65 per cent of total compliance costs, accounted for about two per cent of job losses (Puri 1997, 166). There is also little to suggest that compliance costs in California were substantially higher than those in other states.

In 1992, a consortium of California utilities sponsored the California Industry Migration Study to estimate the number of jobs lost due to relocations out of California and out–of–state expansions of California–based firms. Even with a very expansive definition of jobs 'lost' from California, only 69,000 jobs were lost between 1990 and 1993 because of relocation or expansion outside of California. This is 11 per cent of the 600,000 jobs that disappeared from the California economy in this period. This study included California manufacturing relocations and expansions forgone, for all reasons. Thus, jobs lost due primarily to environmental regulation during this time period would be less than 11 per cent.

Contrasting the surveys of California business people and with the quantitative economic data, Anil Puri concluded:

> [t]here is a clear dichotomy between businesses' perception of cost and the actual cost of air quality regulations. While firm managers are clearly angry at government regulations and view them as costly to business, there's little reliable, quantitative data to support the notion that heavy costs are imposed on the economy by air quality regulations (1997, 166).

And yet, these perceptions persist.

Empirical Studies on New Plant Location

Although it seems impossible to persuade politicians of this fact, there is little state and local governments can do to divert new investment to their jurisdictions. In particular, there is little governments can do in the short term. In the longer term, they can change their comparative advantage by investing in education and infrastructure. In the short term, there is little governments can do to substantially change labour cost or quality. Right–to–work legislation is one of the few policies under state government control found to have an independent effect on new plant location (Holmes 1998).

Unless we treat right–to–work legislation as a type of regulation, there appears to be no area of regulatory policy that offers leverage in attracting new investment. Yet the perception persists that differences across states on these factors, such as the cost of workers' compensation, harm or benefit individual states. An early study of variation in workers' compensation concluded:

> [our studies] deprecated the possibility that these cost differences could cause interstate movements of employers. Nonetheless, some states fear that such cost differences can drive employers elsewhere. Reforms in state programs, which will lead to higher insurance costs, are sometimes avoided because of the specter of the vanishing employer, even if the apparition is a product of fancy and not fact (Watkins and Burton Jr. 1973).

A 1995 study by one of the same authors found that, between the regions, there was a difference of only eight cents per employee–hour in workers' compensation costs (Burton Jr. 1995). Reviewing these findings, a somewhat exasperated economist concluded: '[b]ut the issue refuses to die, and Burton has subsequently made a career of estimating the cost of workers' compensation insurance to employers' (Hunt 1997, 105).

New Plant Location and Environmental Regulation in the US

The issue of environmental regulation and new investment also refuses to die. Robert Tannenwald (1997) reviewed the econometric literature on regulation and new plant creation. The largest body of such studies focused on manufacturing in the US, using state or county level plant establishment data. In general, the studies found that the stringency of environmental regulations does not have generalized effect on new investment across jurisdictions. Economist Wayne B. Gray (1997) found that, where there are negative effects on new investment, they are on the same order of magnitude as the rate of unionization in the jurisdiction. It appears that, to date, no studies have simultaneously compared the effect of environmental regulations against other factors such as tax rates or right–to–work laws.

Researchers assessing Germany's long term viability as an industrial power found that Germany's lengthy permitting processes were more of an obstacle to investment than the level of its regulations, which are about as stringent as US regulations (Blazejczak and Löbbe 1993; Dose et al. 1994). When surveyed, American site selection professionals also express concern about the timeliness and complexity of permitting (Lyne 1990, 1134).

Materials for Site Selection Professionals

If the differences in environmental regulations are important in choosing locations for new plants, this should be reflected in the texts used to teach business location and in the information targeted to siting professionals. A survey of texts, periodicals and websites shows that differences in environmental regulations receive little attention.

One major text on site selection mentions environmental regulations in passing a few times (Harrington and Warf 1995). A perusal of *Site Selection* magazine, the most important periodical for the site selection industry, finds little discussion of differences in environmental regulation across jurisdictions. In a *Site Selection* survey of corporate real estate executives, only 10 per cent of respondents indicated that their company had relocated out of areas where environmental regulations has become particularly tough (Lyne 1990, 1134). *Site Selection*'s articles on the environment focus on model companies and practical information on ensuring compliance (for example Lyne 1991, 586; Zyma 1991, 1082). Location factors mentioned in issue after issue are: cost and quality of local labour force, distance to major markets, infrastructure (particularly highways), taxes, quality of life issues and right–to–work legislation.

As will be discussed in the Canadian chapter, provincial politicians in Canada are very concerned that environmental regulations harm their competitiveness. All available evidence suggests that Canadian environmental regulations are less stringent than those in the United States. (For example, there is no Canadian equivalent to *Superfund* legislation, which imposes legal liability for remediation of old toxic sites.) In addition, the sparse Canadian data on compliance and enforcement suggest less enforcement and lower levels of compliance in Canada than in the US. Yet these facts are not mentioned in the provinces' advertising to attract new investment. None of the provinces' advertisements in *Site Selection* mentioned environmental regulations.

Is Trade Leading to Pollution Havens and Capital Flight?

There is widespread concern about capital flight to jurisdictions with lower environmental standards, particularly in developing countries. Increasing trade flows might accelerate this process. While there is certainly anecdotal evidence of capital flight to pollution havens, it is not clear that relocations are due to environmental regulations. It may be difficult to isolate the impact of environmental policies from wage costs, which for almost all industries, are a far greater cost of production than pollution control.

Econometric studies do not find a systematic shift in the pollution intensity of exports from developed countries or of imports from developing countries. Studies looking for evidence of capital flight from developed countries or to pollution havens have generally not found it. Studies looking at particular industries conclude that new investment and increased production in developing countries result from greater demand in those countries, not exporting into developed countries (Heerings 1993). Surveying the literature, economist Eric Neumayer (2001, 55) concluded 'the evidence for pollution havens is relatively weak at best and inconclusive or even negative at worst'.

From the preceding discussion, one might conclude that there is nothing to worry about. If governments are unable to improve their economic performance by cutting environmental protection, then governments will set optimal environmental policies. However, it would actually be better if there *were* clear tradeoffs between growth or jobs and environmental protection: then people would get something for choosing not to regulate pollution. As things stand, people simply get more pollution.

The Puzzle of Perceptions

There is a puzzle concerning the economic impact of environmental regulations: why do people believe (and act on) ideas which have no empirical support? As described above, there is no empirical evidence to support a connection between higher environmental standards and job loss or lost investment. Studying variations in environmental regulations across US states, William R. Lowry (1992, 13) concluded that:

> [t]he argument that these [business] groups use the threat of relocation to alter subnational behavior can take either of two forms. The strong form asserts that businesses do and will relocate to areas where policies match their preferences, thus creating incentives for state policymakers to alter behavior. The milder argument, one of anticipated reactions, suggests that whether or not businesses do relocate, state policymakers *behave as if they might.* In other words, the threat of exit makes the voice of demand even louder. Evidence for the strong argument has been, for the most part, negative. (emphasis mine)

Thus, even though little suggests that firms relocate because of differences in environmental regulations, the issue matters because politicians behave *as though* firms do, or might, relocate on this basis. As Maureen Cropper and Wallace Oates (1992, 694) note in a survey of the environmental economics literature, the absence of evidence for negative economic effects: '... does not preclude the possibility that state and local officials, in *fear* of such effects, will scale down standards for environmental quality'.

The findings from Lowry's statistical analysis of environmental protection across state governments are consistent with fears of industry exit by state governments. Lowry finds that state policies are not specifically tailored to the severity of pollution problems in that state, thus these policies are not optimal. In addition, he finds that state environmental policies are significantly affected by horizontal competition. Specifically, forms of pollution which are more sensitive to the threat of industry exit (stationary source air pollution and point source water pollution), show less state level innovation. In these policies areas, states are reluctant to exceed minimum federal standards. In the absence of strong federal oversight, state policies in these policy areas show wide variability. This is what we would expect to observe if states' regulations are affected by firms' possibilities for exit.

Adam Jaffe et al. (1995, 148) note that important pieces of US environmental legislation, such as the *Clean Air Act* amendments, were justified by claims that they would prevent interstate competition on the basis of environmental regulations to attract investment. The authors argue that the observed behavior of legislators is consistent with the belief that regulations affect investment, trade and employment:

> There appears to be a widespread belief that environmental regulations have a significant effect on the siting of new plants in the United States. The public comments and private actions of legislators and lobbyists, for example, certainly indicate that they believe that environmental regulations affect plant location choices ... the evidence from the US study suggests that these concerns may not be well founded.

These issues play out in the same way in other locales. An increased role for the European Union in environmental policy was justified by the need to equalize the conditions of competition. For example, the preamble to Directive 84/360/EEC on the Combating of Air Pollution from Industrial Plants states: '[w]hereas the disparities between provisions concerning the combating of air pollution from industrial installations ... are liable to create unequal conditions of competition and thus have a direct effect on the functioning of the common market ...' The European Commission employs this justification even for pollution problems without transboundary implications, such as clean beaches or drinking water standards. Studying the situation in Canada, Kathryn Harrison (1996) has found that fear of losing investment to other provinces was a considerable deterrent to unilateral action by Canadian provinces in regulating the pulp and paper industry.

There is considerable evidence that politicians behave as though this theory were true:

> ... there is a conflict between the anecdotal evidence from legislators and lobbyists, who act as though environmental regulations alter industry location choice and investment, and the empirical evidence presented by economists who

have yet to find systematic statistical evidence of this effect (Levinson 1993, 42).

Unfortunately, politicians, bureaucrats and lobbyists have not been systematically polled on this point. There is not much survey research of legislators and bureaucrats on economic development policies in general. One study of state legislators and bureaucrats found that both groups strongly supported aggressive pursuit of new manufacturing plants (Carroll et al. 1987).

One of the few surveys of policymakers on the issue polled American state regulators and legislators (Engel 1997). The study reported that 88 per cent of state regulators identified industry relocation as an issue in environmental decision–making in their state. As a group, regulators ranked environmental stringency last among five factors affecting industry location decisions. However, the same survey found that legislators ascribed far more importance to environmental standards and education standards than regulators did. There are limitations to this survey, however. The total sample size was only 200 people and the response rate for legislators was particularly low, at 24 per cent (Engel 1997, 378). On the other hand, there is no evidence, anecdotal or otherwise, that the perceptions of politicians coincide with those of economists.

The perceptions of bureaucrats and legislators probably differ more on environmental regulations than on economic development policies because of differences in bureaucrats' incentives for these two types of policies. Economic development bureaucrats tend to push for aggressive policies to woo firms because their jobs would have no justification otherwise (Buchholz 1998, 225–65). By contrast, there is little rationale for employing environmental bureaucrats if *laissez–faire* environmental policies are pursued.

Hypotheses on Legislators' Perceptions and Actions

The discrepancy between the optimal environmental standards governments should ideally set, in theory, and those they actually do set, has received little study. There has been even less consideration of the factors which could account for this discrepancy. David Buchholz's work (1998) on industrial location incentives offers a few hypotheses. These hypotheses, however cannot account for systematic differences between state and national level regulation.

Symbolic politics: 'Do something!'   Politicians' behavior may be due to symbolic politics, the need to appear to be doing something. Buchholz found that politicians' support for incentives derived from the need to be seen to be doing something to attract investment and create jobs. Elected officials were strongly motivated by the desire to point to new investments or to prevent the closure of existing firms. They were more interested in ribbon cutting than long term costs and benefits of having 'done something' to promote or protect the local economy (Buchholz 1998, 274). The factors that are important for location decisions are wholly or partly outside the control of local politicians: distance from major markets, natural resources endowments and wage rates. In contrast, environmental regulations and taxes are factors over which politicians may have some discretion.

Symbolic politics: 'Favourable business climate' Buchholz also found that politicians favoured incentives as a signalling device, indicating a favourable business climate in their city or state. He cites a study of tax incentives which found that New York state legislators thought that incentives were futile but that they were a symbol of a good business climate (Pomp 1988). It is plausible that jurisdictions would want to appear flexible, if not actually lax, on environmental regulations for the same reasons.

Symbolic politics: Ideology Ideology may play a role. For some politicians, limited environmental protection is a core political value. Opposition to environmental protection corresponds to limited government, as well as the defence of private property and free enterprise. Some states, such as North Carolina, have signaled a credible commitment to limited environmental protection by prohibiting state regulations more stringent than the federal minimum (Lowry 1992, 77).

Time horizons Time horizons may be another factor. Buchholz found that politicians' willingness to employ incentives was a function of their ability to shift costs into the future. This is particularly true of tax abatements, which run many years into the future. Many environmental problems also take a long time to show effects. For example, it takes decades for carcinogens accumulating in the environment to cause cancer in humans. Politicians would have few incentives to incur costs in the present to prevent future costs so far in the future.

Conclusion

Concerns about competition and competitiveness, although probably baseless, still weigh heavily on the minds of politicians and regulators in making environmental policy. These concerns affect politicians' actions, without regard to the extent of pollution spillovers to other jurisdictions. This suggests, that contrary to the theoretical economic models, local governments do not set optimal environmental standards, even in the absence of pollution spillovers into other jurisdictions. In the presence of spillovers, these policies are even less optimal for the whole. While this pattern of suboptimal standard setting is at odds with the standard treatment in economic theory, it is analogous to the suboptimal outcomes generated by interjurisdictional competition in incentives to attract investment.

Chapter 4

The Alternatives to Decentralization: Contracts versus Institutions

Chapter 2 discussed how externalities are one of the strongest arguments against decentralization of environmental policy. Does this mean that the existence of externalities is a sufficient condition for centralization? Those who favour limited government argue that externalities are not a sufficient condition (Buchanan 1962).[1] They claim that negotiated agreements and cooperation can produce socially optimal outcomes, without central government action. This study finds that centralized or hierarchical approaches to environmental policymaking systematically result in higher levels of environmental protection than decentralized approaches. The resulting level of protection may, hypothetically, be excessive for social welfare. However, there are many cases where a cooperative or totally decentralized approach results in suboptimally low levels of protection. Although the level of environmental protection in a centralized system may be too high, it is clear that more centralized federations have been relatively more effective in reducing pollution. Centralized federations, which function as hierarchies, are able to overcome obstacles, which inhibit effective cooperative agreement. This chapter discusses just what those obstacles might be.

What can we predict about environmental cooperation? When is it likely to occur and to be effective? A key assumption of international relations theories of cooperation is that parties cooperate when they stand to benefit from cooperation. Cooperation is self–interested, not altruistic. But opportunities for cooperation are often foregone. A common reason for missed opportunities for cooperation is collective action problems, particularly fears of defection from cooperation. Unless there is a credible prospect that cheaters will be punished, cooperation cannot be sustained. It may not even be able to start. These may not be the only problems, however. Credible enforcement does not address bargaining problems, the question of how to distribute gains from cooperation.

This chapter begins by situating cooperation and hierarchy within transaction cost economics. The next section contrasts different types of goods, followed by a discussion of two types of environmental externalities: unidirectional and reciprocal. Next, these technical externalities are contrasted with pecuniary externalities, which work through the price system and do not represent market failure. The fourth section presents the Coasian solution to unidirectional pollution, surveying the theoretical and empirical literature on the question. The fifth section examines

---

1   James Buchanan argues that governments are as likely to impose externalities as to correct them.

cooperation between subnational units within federations. The sixth section introduces the theoretical foundations of international cooperation, beginning with Axelrod's findings on sustaining cooperation in repeated Prisoner's Dilemmas. The seventh section examines claims made for international regimes as vehicles for eliciting cooperation. The final section hypothesizes why hierarchical governance is more effective than cooperation in producing environmental protection.

Transaction Cost Economics

At the most rarefied level of abstraction, arguments about contracts, cooperation and hierarchy are based on the concept of transaction costs. The *New Palgrave Dictionary of Economics* (Eatwell et al. 1987, 56) defines transaction costs as:

> those [costs] of information, negotiation, drawing up and enforcing contracts, of delineating and policing property rights, of monitoring performance, and of changing institutional arrangements. In short, they comprise all those costs not directly incurred in the physical process of production.

Economist Ronald Coase (1960) argued that under certain conditions, an optimal solution to an externality problem would emerge from negotiation between the polluter and the victim of pollution (the pollutee).

Coase received the Nobel Memorial Prize in Economics for this article and one other, 'The Nature of the Firm' (1937). Coase's insight here was to ask: why do firms exist? If contracts are enforceable, why do we not simply observe a nexus of contracts, perhaps like a contemporary movie production? When we look around, why do we see *organizations*, and particularly, ones with a hierarchical structure of command? The common thread in both of Coase's article was transaction costs. His optimal bargain on pollution would emerge in the absence of transaction costs, if property rights were clearly identified. For the second article, the existence of transaction costs explains why firms exist. The cost of negotiating and enforcing contracts makes it cheaper to organize many economic activities in the hierarchical entity of the firm.

Oliver Williamson's contribution to understanding transaction costs was to examine the relative magnitude of these costs in different governance arrangements. Williamson (1989, 142) defined the question thus: '[t]ransaction costs analysis entails an examination of the comparative costs of planning, adapting and monitoring task completion under alternative governance structures'. He classified transactions on the basis of three parameters:

- frequency
- uncertainty
- asset specificity (the most important).

Williamson emphasized how these parameters affect opportunism by players, the likelihood of being cheated in a transaction. Thus as transactions become less frequent, more uncertain or higher in asset specificity, the risk of opportunistic behavior by the other party increases. Williamson also argued that, under conditions of information asymmetry, a small number of players available to contract with makes opportunism more likely (1975, 9). A large number of potential contract partners more closely resembles a market, and as a result there are fewer incentives for strategic behavior and more information available on prices.

With regard to international cooperation, international regimes are thought to reduce many transaction costs (Keohane 1983). The literature has tended to look at transaction costs as a whole, not distinguishing transaction costs due to enforcement from others. A comparative analysis of cooperation in federal systems can be helpful in this analysis. Because enforcement problems are more manageable within an established legal system, it is possible to focus on other potential obstacles to cooperation, such as fear of opportunism or distribution of gains from cooperation.

Table 4.1 Types of goods

		Rivalness of consumption (extent to which consumption of the good by one person affects consumption of the same good by another)	
		none	**complete**
Excludability of consumption (ability to make users pay for the good)	**none** (physically impossible or prohibitively expensive to exclude)	public goods (*such as national defence or air quality*)	common property goods (*such as an ocean fishery*)
	complete	joint goods (*such as satellite TV programming*)	private goods (*such as a sandwich*)

Externalities, Including Public Goods

Externalities encompass a broad range of market failures. Externalities exist when the production or consumption of a good affects the welfare of others, outside of the transaction, in ways that are not accurately reflected in prices. They can be beneficial, such as basic research or harmful, such as pollution. Public goods are one type of externality (Table 4.1).

Public goods (and bads) are classes of externalities. Water quality and air quality are public goods. Water can also be a private or common property good, depending on how it is used. It is possible for a city to prevent residents from consuming water unless they pay their water bills. However, it is not possible for one jurisdiction to prevent another from consuming water, unless that jurisdiction is upstream and diverts a substantial amount of water out of the river preventing consumption by downstream communities.

When used for waste disposal, water and air are public goods. In practice, governments have found it difficult to exclude firms and individuals from dumping waste into air and water. The ability of the air and water to absorb wastes is subject to some congestion but that is under fairly extreme conditions. The ability of the air or water to carry off wastes is not significantly different for the fifth smokestack or sewer pipe than for the fiftieth. For the 500th emitter, however, congestion may begin to set in. However, while the air or water's utility for waste disposal is unaffected, the consequences for other uses can be disastrous. The air's usefulness for breathing or the water's utility for drinking or recreation is seriously degraded as pollution increases.

There are two types of transboundary environmental externality: non–reciprocal and reciprocal. In theory, the type of externality has implications for the possibility of reducing the externality cooperatively or non–cooperatively. Nonreciprocal externalities correspond to pollution on a River; reciprocal pollution is analogous to the situation on a Lake. Cooperative agreements between governments are usually proposed as the solution to River and Lake–type problems. Coasian bargains have been proposed as a solution to River–type problems.

The River: Nonreciprocal

Picture a river with a series of towns on its banks as it runs to the sea. All the towns dump their sewage into the river and draw their drinking water from the river. Water pollution in the River is a case of a non–reciprocal (unidirectional) externality. If each town acts unilaterally (non–cooperatively), no town has an incentive to treat the sewage it produces (Barrett 1991, 166–7). The town the furthest upriver, whose pollution reaches all the other towns, has no incentive because the River is unpolluted when it reaches this town. The towns downstream, receiving the River's water in a progressively more polluted state, have no incentive to treat their sewage because the quality of their drinking water depends, not on their own sewage, but that of all the upstream communities. The River is analogous to air pollution problems where there are prevailing winds that transport air pollution in one direction, always in the same direction.

The Lake: Reciprocal

The Lake is the model of a reciprocal externality. Imagine several communities around a lake, all of which draw their drinking water from the Lake and discharge their sewage into it. This scenario is reciprocal because each community is affected by the pollution from all the communities, including its own pollution. Under non–cooperative conditions, where each town acts unilaterally, each town has some incentive to abate its own pollution (Barrett 1991, 141). The size of that incentive depends on the population of the town and the total number of towns. If there is one large city and a few small villages on the Lake, the city will face the same incentive for sewage treatment as if it were the only town on the Lake. If, however, there are dozens of communities of roughly equal size, incentives for unilateral pollution abatement are much smaller. Because each town has *some* incentive to abate its own pollution, cooperation should be more likely than for the River. The Lake is analogous to air pollution in a basin, such as the Los Angeles Basin, where little pollution is imported or exported.

To summarize, we do not expect to see unilateral action or cooperative agreements to water pollution on the River, because there is no incentive to abate one's own pollution. The downstream communities might have an incentive to pay the upstream communities to reduce their pollution. In contrast, communities on the Lake have some incentive to abate their pollution unilaterally and even more to cooperate in abating their pollution. But this analysis deals only with the externality created by the pollution.

The Role of Pecuniary Externalities and Cooperative Agreement

There are also pecuniary externalities. These are the effects on prices paid by one firm or consumer on the prices paid by other participants in the market (Shubik 1984, 407). For example, the lemonade stand that opens up next to yours does not create an externality in the strict sense, but if it forces you to drop your prices, it has created a pecuniary externality. Classical externalities, both positive and negative, such as public goods and pollution, are types of market failure. Market failure is a justification for government intervention. However, pecuniary externalities are *not* a class of market failure. Much to the dismay of failing firms and sectors, a pecuniary externality *does not* justify government intervention. Pecuniary externalities simply reflect the operation of the price mechanism.

If we assume that a government will act rationally to balance costs and benefits to its jurisdiction, it should be easier to cooperate on Lake pollution abatement than on River pollution, because of the greater likelihood of capturing benefits for oneself. If, however, governments are extremely concerned about the cost of abatement and its impact on the competitive position of their town, cooperation may not be any easier on the Lake than on the River. As noted in the previous chapter, it appears that governments significantly overestimate the price effect of pollution abatement in their jurisdiction.

Experience in the US during the period of interstate compacts found that cooperative solutions to water pollution on lakes were *not* easier to negotiate nor more effective than those for rivers. In fact, the most effective water pollution control compacts in the US all involved rivers.

Historical experience in the US and Switzerland indicates that pecuniary externalities have played a significant role in cooperative pollution abatement. In the case of a River which empties into the ocean, it is far more important for the community farthest upstream to treat its sewage, than for the one on the ocean to do so. Yet, in cooperative settings or those were requirements are centrally imposed, abatement requirements are invariably *the same* for both communities. In Switzerland, when governments reached cooperative agreement to require sewage treatment, all parties were adamant that the same treatment requirements should apply to all, without regard to the impact of their pollution (or the cost). Economists consider this outcome to be highly inefficient because this does not maximize the environmental protection for the infrastructure investment.

The norm seems to be 'if we have to do this, we should all be harmed equally!' Communities seem concerned with this particularly rigid norm of equity: an equal level of treatment required by all. However, requiring the same level of treatment by all parties is not the same as imposing the *same costs* on all. In one respect, when a higher order of government imposes a uniform standard of treatment on all, and finances all sewage treatment out of its revenues, this does have the effect of equalizing the cost imposed. Economists judge this outcome to be highly inefficient. However, we do not observe this outcome because even very generous federal financing always requires some level of local matching funds. The unit of cost of sewage treatment will thus vary by community based on economies of scale and the local tax base. This is still suboptimal from an economists' point of view because it does not allocate abatement resources according to the severity of pollution effects. In the cooperative situation, governments do not appear to care about overall economic efficiency – getting the most pollution abatement for a particular investment of society's resources. Governments wish to spread the economic disadvantage around.

Taken to the (il)logical extreme, the concern over pecuniary externalities means that jurisdictions will not act unilaterally to abate pollution that affects no one but themselves. This is analogous to refusing to brush your teeth, unless your neighbours brush theirs. In an individual's cost/benefit calculation over tooth brushing, the actions of one's neighbours should be irrelevant. Who cares if your neigbour gains an extra five minutes a day over you? Will having cavities feel better, known your neighbours have them too?

In the context of American interstate environmental compacts, one of the obstacles to negotiating an effective compact was the fear that the participants would be at an economic disadvantage *vis–à–vis* everyone *not* in the compact. This line of thinking was common, even in the case of lake compacts where *all* the benefits are captured by the participants. Imagine the problems this mindset can create in cases where the participants to the deal cannot expect to capture all of the benefits, such as the Kyoto Protocol or other agreements on climate change.

The Coasian Prescription for Pollution Control

Coasian bargains over pollution are much analyzed in theory but virtually never observed in the real world. Coase argued that in the presence of well–defined property rights, producers of pollution and those suffering its ill effects could reach a bargain over side–payments. The end result would be an efficient level of pollution. The Coase Theorem holds that an efficient solution to a pollution problem does not depend on the initial assignment of property rights. For efficient resource allocation, it does not matter whether people suffering ill effects from a polluting factory pay the factory to reduce its pollution or if the firm reduces its pollution in order to minimize damages paid to those who are suffering from pollution. If the people suffering the ill effects think it is worth the money, they will pay the factory owner to pollute less. The pollutees will pay the money because, unless they are made better off, they would not enter into an agreement. In that case, the outcome is also Pareto improving.

The literature finds few examples of Coasian bargains over pollution, whether between governments or private parties. A few examples of Coasian pollution bargains have been identified in the literature, although none correspond exactly to the bargain described in Coase's article. The existing examples of Coasian pollution bargains between governments are all at the international level, not within federations. Thus far, they have all been River–type pollution (unidirectional externality). Economist Scott Barrett (1991, 166–7), who models international environmental cooperation, argues that '... agreements dealing with unidirectional externalities are rare and almost invariably toothless'.

The first case concerned the use of financial transfers to reduce chloride pollution in the Rhine, which disproportionately affects waterworks and farmers in the Netherlands. The largest single source of pollution was government owned potash mines in France. The governments of France, Germany and Switzerland, as well as the Netherlands, made payments to reduce chloride pollution from the mines. Thomas Bernauer found that, contrary to expectations, the Coasian bargain to the problem of chloride pollution was inadequate: 'too little, too late' (1996). He argued that the protracted negotiations and unsatisfactory outcome were the product of information asymmetries and strategic behavior on the part of several players. For example, Germany reached agreement with the French government to reduce pollution from the mines that jeopardized groundwater in Germany. However, elimination of this threat resulted in increased chloride pollution in the Rhine River (Bernauer 1996, 222).

In the second case, Japan tried to use aid and technology transfer to prevent air pollution from China (Evans 1999). China has lots of cheap coal, which it is using to meet growing demand for electricity. Increased coal consumption by China harms air quality in Japan and Korea, as well as in China. Japan sought to encourage better combustion technology in Chinese power plants, to reduce total emissions. In studying the arrangement, Peter Evans characterized it as an arrangement that reduced Japan to pleading with China, without securing any guarantees of maintaining air quality.

The third case was a deal between a French town and a German one, on either side of the border. The German town paid the French town to *not* build an incinerator (Feld et al. 1996). The case is even more intriguing because the funds were raised by voluntary contribution, not taxation in the German town, which suggests that the free–riding problem may not be insurmountable under decentralized conditions. The German residents developed an alternate project for the derelict land where the incinerator was to be built and paid to restore the land. The French town sold the land to the developer for a nominal sum of one Franc, because the French community also stood to gain from a less polluting land use.

This case differs from Coase's example because those making the payment are less vulnerable to opportunism, because the negotiation resulted in a source of pollution not being built. By subsidizing an alternate use of the land, the German residents obtained a credible commitment that the land would not be used for an incinerator. In Coase's example, the pollutees pay the polluter to reduce pollution, not shut down the plant. In Coase's example, the pollutees are more vulnerable to opportunism. If they make a lump sum payment for a certain amount of pollution reduction, they cannot be sure the producer will not threaten return to the old level of pollution, unless more payments are made. In theory, the Germans could have paid the French town to build a less polluting incinerator. The German residents' concerns were motivated by fear of a drop in property values. It is not clear that a somewhat cleaner incinerator would have had less negative impact on property values.

Several arguments have been offered to explain the paucity of Coasian bargains in the real world. We can take Coase literally, and argue that we do not observe Coasian bargains because transaction costs are never zero in the real world. Robert Keohane (1984, 87) argues that Coase's assumptions of clear property rights, perfect information and zero transaction costs are not found in the real world. If Coase's conditions were fulfilled, international regimes would have no value–added and hence would not exist (Keohane 1983, 154).

Other commentators take issue with Keohane's characterization. First, they argue that zero transaction costs are not a necessary condition for Coasian bargains to emerge. According to John Conybeare, the clear assignment of property rights is not a problem in the international sphere. The principle of national sovereignty in effect creates a liability regime of 'victim pays'. Conybeare (1980) argued that Coasian bargains are better suited to resolving public goods problems in the international system than international organizations are.

Perhaps information is the more significant obstacle. Joseph Farrell (1987, 115) argued that the existence of private information reduces the usefulness of the Coasian framework, drawing upon research showing that deals are quickly negotiated when the bargainers' payoffs are common knowledge. Farrell's model found that, in the presence of private information, Coasian bargains are not unambiguously preferable to centralized action by a 'bumbling bureaucrat'.

One major transaction cost, perhaps the largest, is the cost of enforcing agreements. This cost should be less within federal systems than between countries at the international level. Subnational governments within federations have access to courts with powers to enforce agreements. In addition, within English common law federations, such as Canada and the US, property rights, at least with regard to water, are clearly defined. The doctrine of riparian rights dictates that downstream water users have the right to receive water in its natural state. They should not have the condition of their water or the flow of it, affected by the actions of upstream users. Downstream owners can seek relief from the civil courts, in the form of an injunction and damages (Ball and Bell 1991, 457). The fact that we do not observe Coasian bargains within federal systems suggests that other obstacles must be at work.

Other commentators have focused on obstacles not explicitly identified by Coase. Karl-Göran Mäler (1991) emphasizes reputation effects, particularly when bargaining over other issues or with other parties. He argues that downstream countries are very reluctant to accept the 'victim pays' principle, because they are fearful of getting a general reputation for weakness. Mäler and others also note that, in the case of multiple downstream countries, there are collective action problems in negotiating a Coasian bargain. At least in the short run, countries have an incentive to free ride, hoping that another downstream country will pay for the pollution reduction (Mäler 1991, 191).

Robert Cooter (1982) raised a more general objection, focusing on the problem of splitting gains. He argued that, while the existence of mutual gains is necessary for cooperation, the existence of those gains offers no solution to the problem of dividing the gains between participants. The problem goes beyond transaction costs: it is a bargaining problem. Thus, mutually beneficial bargains will remain unrealized; money will be left on the table, metaphorically speaking. The extent to which division of gains is a problem will depend on how participants value absolute versus relative gains from the bargain.

Examining the Coasian bargain over chloride pollution in the Rhine, Thomas Bernauer identified a puzzle. This Coasian deal between wealthy governments with high administrative capacity, under favourable contracting conditions was strikingly ineffective and took a very long time to reach. There is a need for more research to identify factors that facilitate or inhibit Coasian bargains. If transaction costs are the problem, which are most significant? Are they enforcement costs? Information costs? For many examples in the literature, such as international cases, the outcome of no Coasian bargain is overdetermined, because several problems are present.

Cooperation: Within Federations

Although Coasian bargains do not appear to occur in federations, formal cooperation within federations does have a history, albeit a neglected one. Formal cooperation among the American states has received little attention from scholars and practitioners. The situation is reflected in the title of the only recent book devoted to the subject: *Interstate Relations: The Neglected Dimension of Federalism* (Zimmerman 1996). This work is wholly descriptive.

While now neglected, interstate cooperation was once a very hot topic. In the 1920s and 1930s, interstate compacts were seen as *the* solution to problems which were too big for the states to handle but which the federal government could not or should not address. Until the United States Supreme Court decision of 1938, upholding the constitutionality of the New Deal, the American federal government was unable to tackle a wide range of social and economic problems because it lacked jurisdiction.

Interstate cooperation is ripe for re–evaluation. If American conservatives such as the Federalist Society have their way, American state sovereignty will once again come to the fore. Economists Robert Inman and Daniel Rubinfeld (1997 60) have argued that: '... recent fiscal reforms in welfare policy appear to be a significant institutional experiment with an alternate paradigm of federalism, one which emphasizes the ability of states, not the central government, to handle cross-jurisdiction spillovers'.

If American states gain powers and responsibilities, deep cooperation between the states may once again become an issue. Many mechanisms are in place for shallow or informal cooperation, such as information sharing. Groups such as the National Governor's Association serve as information clearinghouses but in no way constrain the behavior of its member state governments. In the US and Switzerland, mechanisms for formal, legally binding cooperation exist but have been overlooked for many years.

In the 1920s and 1930s, there was tremendous optimism about the prospects for interstate compacts. The best known advocates were legal scholars Felix Frankfurter and James Landis (1925). Many books and articles were written, extolling the promise of this form of governance (see Fite 1932; Clark 1938). Compacts were also seen as a means to achieve more holistic and authentic government, rooted in the distinctiveness and diversity of American regions rather than the individual states or uniform federal measures (Frankfurter and Landis 1925).

Frankfurter and Landis (1925) proposed compacts as the solution to problems that generated interstate externalities, such as navigation, fisheries management, utilities regulation and taxation for example. Ironically, these are *precisely* the kinds of problems which compacts failed to resolve or where negotiations were never successfully concluded. The compacts which were highly successful and persist into the twenty–first century are those without distributional consequences. These include compacts with reciprocity, such as mutual recognition of driver's licenses or reciprocal supervision of parolees. Compacts that coordinate economic activity also gained a large number of signatories within a few years, for example, compacts for nation–wide uniformity in commercial transactions.

Frankfurter and Landis (1925) argued that compacts were well suited to resolving interstate disputes. They argued that the Supreme Court was unable to promptly resolve problems such as interstate disputes over water apportionment or sewage treatment in the greater New York area but compacts had resolved the impasse. What they failed to recognize was that the compacts to address these problems, while not imposed, arose in the *shadow* of hierarchy, the hierarchy in this case being US Supreme Court decisions. Fritz Scharpf (1993, 13) has argued that the threat of hierarchical intervention can contain opportunism: 'while, in the typical case, outcomes are negotiated, rather than hierarchically imposed, the success of these negotiations is greatly facilitated by the fact that they are conducted "in the shadow of hierarchical authority"'.

The knowledge that the alternative to a voluntary agreement is further Supreme Court litigation restrains opportunistic behavior and encourages cooperation through compacts. Although Supreme Court cases proceeded slowly, the states in these settings did not hesitate to go the Supreme Court, sometimes repeatedly. For example, the downstream states in the Delaware River basin took New York State to the Supreme Court in 1929 and in 1954. The outcomes of compact negotiations represent a mutually agreed upon improvement on the likely outcome of Supreme Court decisions. Thus these cases cannot be treated as success stories for compacts as a means of independent conflict resolution.

Two contemporaries of Frankfurter and Landis asked:

> [i]s there a middle road between the states and the national government? ... [W]ell-established interstate compacts are put forward as a possible alternative to impotent, devitalized states on the one hand and an overburdened national government on the other. Can interstate compacts deal satisfactorily with problems that are too large for the separate states or beyond the power of the federal government? ... Are present hopes well-founded or are the possibilities of interstate cooperation something of a delusion? (Dimock and Benson 1937, iii).

These authors concluded the compacts were most likely to succeed for problems that had no element of interstate rivalry or competition. The corollary was that compacts were unlikely to be of any use in situations of serious conflict of interest between states. In contrast to Frankfurter and Landis who wrote in 1925, these authors writing in 1937 had the benefit of 12 years worth of hindsight on compacts. They were also well acquainted with the intractable problems that the Great Depression created for all levels of government.

There have been cheerleaders for noncentralization and interstate cooperation in the latter half of the twentieth century as well. Morton Grodzins (1966) pointed to the American national police system and national recreation policy, and praised 'the virtues of chaos'. Advocates such as Daniel Elazar and Joseph F. Zimmerman did not assess the efficacy of cooperative arrangements in terms of problem solving capacity. Nor have of the advocates have provided an account of the assumptions underlying this form of governance: none offers an account of *how and why* interstate cooperation is supposed to function. In theory, because interstate agreements are entered into voluntarily, any such arrangement should be Pareto improving. In theory, the benefits to interstate compacts might include:

- economies of scale
- internalizing externalities within a region
- more opportunities of learning from experiences in other jurisdictions

However, the existence of mutual gains, while a necessary condition for cooperation, is not sufficient. It seems just as plausible that Prisoner's Dilemmas or other collective action problems could result.

Albert Breton notes that the study of cooperation in federal systems has focused on vertical cooperation, between the federal government and the states. Horizontal cooperation, cooperation among units at the same level of jurisdiction, has received far less attention. He argues that cooperation is not a solution to problems of competition between subnational governments because neither interstate competition nor cooperation is stable. Breton (1996, 249) stated: '[d]iscussions of cooperative federalism are mostly concerned with vertical relationships: they are largely tacit on the question of horizontal competition. The point is that horizontal cooperation is not a solution to horizontal competitive instability'.

Cooperation: At the International Level

If the federalism literature makes little use of the international relations literature on cooperation, the converse is also true. In the international relations literature, little use has been made of natural experiments in federal systems to test international relations theory. Daniel Deudney's work (1995) is the exception that proves the rule: between 1787 and 1861, the US was what Deudney calls a 'Negarchy': neither anarchic nor hierarchic.

The great advantage of studying cooperation in federal systems is that most of these systems have credible enforcement capacity. This makes it possible to control for fear of defection, because those fears should be assuaged, at least relative to the international context. Formal cooperation in federations suggests that enforcement capacity is not sufficient to generate cooperative agreements. It appears that issues of relative gains, including those that might go to third parties not in the agreement, seriously inhibit cooperative agreement. Concerns about distribution cannot be overcome through effective enforcement capacity.

What do we know about cooperation in the absence of enforcement capacity? Robert Axelrod used tournaments to model two–person cooperation in a Prisoner's

Dilemma. He found that sustained cooperation could be elicited in repeat play, if the players used a strategy of direct reciprocity: '... the requisites for cooperation to thrive are that the cooperation be based on reciprocity and that the shadow of the future is important enough to make this reciprocity stable'(Axelrod 1984, 173). The reciprocal strategy players used is Tit–for–Tat: to reward cooperation with cooperation in the next round and to punish defection with defection in the next round. The shadow of the future exists only under conditions of repeat play, where each player *knows* they will face the same opponent again and again.

While Axelrod's result was very significant, it may not support the full weight of the arguments that have been built upon it. His result has been extended to argue that stable cooperation will emerge between more than two players, employing diffuse reciprocity, under condition of repeat interaction (Keohane 1984, 75–6). Critics have focused on several features of Axelrod's model that may limit its applicability to *n*–person dilemmas of cooperation over public goods. First, Axelrod's players face a binary choice: cooperate or defect. This is an accurate representation for many choices in international relations, such as the superpowers' nuclear standoff during the Cold War.

Richard Cornes and Todd Sandler (1986, 141–2) argued that the result Axelrod obtained from Tit–for–Tat:

> ... is certainly interesting, although its relevance to public goods problems is debatable ... even within a binary choice framework (contribute, don't contribute) the public goods problem may not conform to the Prisoner's Dilemma.

Few environmental problems that interest us are characterized by binary choices, although deciding for or against nuclear power or genetically modified crops would be. For public goods, such as air and water quality, choices tend to be continuous variables about how much to cut emissions or how much to contribute. Even assuming a two–person game, how does one punish an inadequate level of contribution?

The principal criticism leveled against the model is that one cannot extrapolate a result from a two–person game to a game with *n*–players. Cornes and Sandler (1986, 141–2) point out: '[i]t is not clear that it makes much sense for an individual to punish or reward 99 others in later plays of the game'. Michael Taylor (1987, 71) argued that the provision of public goods typically involves more than two players and that the decision of a player to cooperate is 'contingent on there being *enough others* cooperating'. Taylor modeled an *n*–person Prisoner's Dilemma and found that cooperation can arise, without regard to the number of players. However, repeat play is not sufficient to guarantee this outcome. Subject to a particular discount rate, cooperation will emerge if each player's cooperation is conditional on the cooperation of all the others. However, the repeated game can also give rise to Chicken, where some players exploit the cooperation of the rest by consistently defecting (Taylor 1987, 104). Thus the emergence of stable cooperation in a *n*–person Prisoner's Dilemma is possible, but hardly certain.

Scholars have looked at existing international institutions and concluded that other international institutions could also be successful. After the successful negotiation of the Montreal Protocol, but before the Rio Summit in 1992, many were optimistic that a climate change treaty could be negotiated as easily as the Montreal Protocol to prevent destruction of the ozone layer. These optimists failed to control for differences in the degree of difficulty in reaching agreement and the nature of the problem.

George Downs et al. (1996) argued that scholars have failed to control for the depth of cooperation found in existing international agreements. Depth of cooperation measures how much the behavior of the signatories departs from what they *would have done* in the absence of agreement.[2] Deep cooperation requires a substantial departure from what the parties would have done otherwise. By this standard, most international agreements require only shallow cooperation, such as sharing information. When it becomes apparent that the negotiating countries cannot agree on an effective solution to a problem, they agree to do what they can agree to. What they agree to do may bear little resemblance to the actions required for a solution to the problem. The agreed upon measure may not be sufficient, but it is the best agreement the parties can arrive at.

Much research on international environmental cooperation has failed to control for the counterfactual: what would have happened in the absence of the regime? If an international regime exists, it is assumed to have had an independent effect on its members' behavior. Much research has focused on counting treaties and signatories to them. There has been much less work on the effectiveness of treaties in improving environmental quality. James Murdoch et al. (1997) conducted econometric tests of the effectiveness of two treaties to improve environmental quality. Their findings support the hypothesis that international environmental treaties are examples of shallow cooperation. They concluded that the Helsinki protocol to reduce sulfur emissions simply formalized reductions that the signatories were going to carry out unilaterally anyway.

Under shallow cooperation, collective action problems such as defection become trivial. If signatories are doing what they would have done anyway, they are merely formalizing unilateral action in a regime. When they behave in this way, they are not behaving strategically: the behavior of one state does not depend on what it expects other states to do. Thus considerations of defection, monitoring and reputation become much less important.

When cooperation is modeled as a Prisoner's Dilemma, the key obstacle to cooperation is defection and the fear of defection by other. Building on Axelrod's work, liberal institutionalist theorists of international relations have identified the following factors as conducive to cooperation: repeated interaction, long shadow of the future, monitoring and enforcement capacity, transparency, multiple issues and opportunity for side payments (Keohane 1984). The single necessary condition for

---

2   An example of fairly deep cooperation is interstate bridge construction. Neither of the parties would build their half of the bridge in the absence of the agreement.

cooperation is the existence of mutual interests or possible joint gains (Axelrod and Keohane 1986). These conditions should foster cooperation because they mitigate the collective action problem. Participants are usually reluctant to cooperate because other participants might defect. The incentive structures generated by international regimes are supposed to overcome the enforcement problems that occur in the absence of a hegemon able to enforce agreements.

Federal systems can represent a critical test of liberal institutionalist theories of international cooperation. Federal systems have all of the conditions that liberal institutionalists consider conducive to international cooperation. Furthermore, all of these factors are present to a greater degree than these are found in relations between countries. Cooperation in federal systems should also be facilitated by the smaller number of actors and the prospect of more concentrated benefits. For certain issues, such as lake pollution, there is a high expectation that the environmental benefits can be limited to a specific group of states. This should make it easier to limit free riding. In airsheds, air pollution has the quality of a Prisoner's Dilemma but with a fixed number of actors. This contrasts with global air pollution phenomena such as climate change, where unilateral action by one region brings only concentrated costs and little likelihood of benefits. Relative to the international system, the smaller number of parties should make it easier to reach agreement.

International regimes have been identified as a mechanism to reduce transaction costs, beyond the cost of enforcing contracts. However, the literature has not examined exactly which transaction costs are reduced by regimes. Nor has Williamson's methodology been followed: comparing different forms of governance to assess their impact on different types of transaction costs. The international relations literature offers a few opportunities for this kind of comparison, particularly comparing periods of hegemony to relatively more anarchic historical periods. Federal systems offer a richer universe of cases to study these questions because there is variation on the governance variable, from informal cooperation, formal cooperation to hierarchical organization.

Within federations we find surprisingly little binding cooperation on the environment, even where legally enforceable agreements can be negotiated. It appears that the capacity to enforce contracts is not a sufficient condition to elicit cooperation. For example, although the Swiss constitution permits formal legal agreements (*Konkordaten*) between the cantons, few environmental agreements are found among the hundreds of concordats signed.

The US chapter focuses on the American experience with interstate cooperation, in the environment and in other areas. There are only a handful of cases of effective interjurisdictional cooperation on the environment in federal systems: water pollution control compacts in the US. Even the successful compacts were far less effective reducing pollution than unilateral action by the US federal government. Most interstate compacts are examples of shallow cooperation: the requirements of the compact differ little from what the signatories would have done anyway, in the absence of the agreement. Compacts could have included monitoring and enforcement, but were not designed to do so.

The cases of successful cooperation involved heavily polluted rivers in the industrialized eastern US. However, not all cases of polluted rivers in the Northeast led to successful cooperation. For example, over decades, attempts to negotiate water pollution control in the New England states came to naught. Pollution problems which led the states to go before the Supreme Court to seek a remedy were more likely to result in effective compacts. One surprising result is that pollution on lakes was *less likely* to be addressed effectively through a compact than was river pollution. The US chapter sets out all the existing and attempted water pollution control compacts, identifying three that seem to have had an independent effect on reducing pollution. The chapter also identifies several watersheds with serious water pollution problems where no compact was ever negotiated. This suggests that cooperation was unable to provide a public good for which there was public demand.

In the case of cooperation through interstate compact, fear of defection has not been the principal obstacle to cooperation. National subunits can prove themselves incapable of co–operation *even when all the subunits stand to gain and gain equally by it*. Side payments or generous grants from the center are not sufficient to produce cooperation. States' aversion to binding themselves, even in agreements of their own making, may be the greater obstacle. Ubiquitous concerns about economic competitiveness are a factor. American states in an interstate compact do not want to disadvantage themselves *vis–à–vis* states outside the compact, in other parts of the country.

Keohane, Haas and Levy (1993, 11) have identified favourable contracting conditions as a prerequisite for international cooperation. Favourable contracting conditions are one of the 'three C's', the other two being a high level of governmental concern and capacity. As noted above, constitutional provisions for interstate compacts in the US (and in Switzerland) were not sufficient to stimulate extensive use of these instruments for cooperative pollution control. The Canadian case lacks the formal constitutional provision of the compact but there is an overarching legal system. The Canadian case shows high levels of concern about the environment in public opinion polls. The provincial governments also have greater fiscal and technical capacity most American state governments. Despite all these factors, provincial governments have been very reluctant to enter into any binding cooperation on the environment, or other matters such as reducing interprovincial trade barriers (Brown 2002). Once again, this suggests that the conditions identified as conducive to cooperation are not sufficient.

Hierarchy versus Cooperation:
Why is Hierarchy so Much More Efffective?

In all cases where federal standard setting was introduced, it proved far more effective than even the best cooperative agreements. There are a few possible reasons why hierarchy outperforms cooperation. First, hierarchies may be more effective than other arrangements at preventing opportunism. Second, centralized federations may generate more effective environmental policies because they are able to overcome distributional problems which block agreement in under cooperation. Hierarchies are able to authoritatively allocate and redistribute resources. The resulting outcome may not be socially optimal, but this form of governance has the capacity to carry out this task, unlike noncentralized or cooperative arrangements.

Williamson argued that hierarchies (or vertical integration) reduced opportunism, particularly for highly specific assets. The owners of transaction specific assets are vulnerable to opportunism when it comes time to renegotiate the deal. The classic example is a factory (B) which has been built close to another factory (A) to serve as a source of inputs. Once the supply Factory B has been built, it may be vulnerable to opportunism. Upon renegotiation, Factory B has no choice but to accept the terms of Factory A, if A can use other suppliers. Traditionally, steel mills at the head of an iron ore or coal mine, tended to be integrated into a single operation to overcome such opportunism. Hierarchical action on an issue like sewage treatment may be easier than cooperative action because there is no fear that the other party will not build its sewage treatment plant as promised or will not put it into operation. This seems like a rather irrational fear however, because the risk in most cases is symmetrical. There is no advantage accruing to one party as a result of the investment, between the time of the initial contract negotiation and a renegotiation.

James Sebenius (1992, 326) argued that much of the international relations literature has gone down a thorny and barren path by focusing on game theoretic models of cooperation, particularly the binary choices of cooperation and defection in the Prisoner's Dilemma. He argued that the negotiation literature shows that cooperation and competition cannot be separated in studying negotiated agreements. Distribution is at the heart of all negotiation: the decision to cooperate in making the pie is cannot be separated from the decision on how to split it. Piemaking is 'value creation' and pie splitting is 'value claiming'. In particular, strategies which are effective for value claiming can be very damaging for value creation. Sebenius (1992, 336) states 'competitive moves to claim value tend to drive out cooperative moves to create it jointly, often leading to Pareto–inferior agreements, deadlocks and conflict spirals'. Hierarchical governance structures can reduce the likelihood of deadlocks by making side–payments.

Conclusion

This chapter has identified a number of conditions which are thought be necessary, although perhaps not sufficient, to generate cooperation. First, the prospect of mutual gain from cooperation is necessary but not sufficient to lead to cooperation to exploit those gains. What is less well known is that the existence of enforcement capacity, as a credible deterrent to cheating and defection, is not sufficient to encourage Coasian bargains or cooperation. No doubt enforcement capacity made a difference in the negotiation of some of the more ambitious interstate compacts, but the existence of courts for dispute resolution still left significant joint gains unexploited, particularly in the case of water pollution on lakes, where interests are roughly symmetrical.

The advantage that hierarchy has over Coasian bargains and cooperation may be the ability to limit opportunistic behavior. It seems more likely that the capacity of hierarchies to authoritatively allocate resources (even if not always) efficiently accounts for the greater efficacy of this form of governance over Coasian bargains and cooperation. It seems likely that conflict over the distribution of gains means that many potentially beneficial deals remain unrealized and that some agreements which are negotiated fall well short of the Pareto frontier of optimal allocations.

Hierarchies may be better able to limit opportunism because enforcement is not endogenous to the agreement. Thus there are limited opportunities to tinker with enforcement in negotiation of the agreement. When enforcement is exogenous, the mechanisms for enforcement are not up for negotiation.

Chapter 5

The United States:
Greening through Centralization

'We can do it ourselves! You just get in our way!' This is the clarion call of the states berating the US federal government, asserting their superior ability to make policy. The 1990s saw a spate of books and articles arguing for greater decentralization of policy, including environmental policy. The claim is that lower levels of jurisdiction can incorporate local knowledge and thus better reflect local preferences and conditions. Where environmental problems have transboundary implications, these problems should be resolved by contracting or by cooperation.

Has the federal role in American environmental protection been superfluous? How well did the US states do when they had exclusive jurisdiction for addressing environmental problems? How well has interstate cooperation worked? Would decentralized environmental policymaking be more representative of public preferences? Would decentralization lead to a race to the bottom or more nuanced state level policies?

This chapter finds that the historical experience of state environmental policymaking is not particularly encouraging. The federal government played a significant role in most improvements in environmental quality, even prior to the federal pre–emption of the late 1960s and early 1970s. More recently, state level action on environmental problems continues to be inhibited by concerns over competitiveness and interjurisdictional competition. In the past, despite numerous attempts that were made to stimulate interstate cooperation, agreements were negotiated in only a few cases. More recent attempts to solve problems with interstate cooperation have not yielded results.

This chapter presents empirical evidence from the United States, from the 1930s to the present. The chapter is divided into six parts. The first parts surveys American experience with noncentralized environmental policymaking for air and water pollution. The second part describes the gradual growth of the federal government's role, from research, financing and facilitation to standard setting and oversight. The third section examines interstate compacts in general, contrasting successful and less successful compacts. The fourth section focuses on the efficacy of interstate cooperation on the environment, up to the present day. The fifth section examines claims made for greater state autonomy in environmental policymaking, in terms of representativeness and adaptation to local conditions. The final section examines policy on drinking water and concludes that the federal government still performs a valuable oversight role.

Historical Evidence: The Inadequacy of Local and State Action

The Era of Local Action: Air Pollution

Prior to the 1940s, the regulation of pollution was seen as a purely local matter. Pollution was controlled by means of municipal ordinances or lawsuits for the abatement of nuisances. Local governments were notoriously unwilling to regulate or prosecute industries, for fear of losing jobs and investment. Pittsburgh, which had some of the America's worst air, is a case in point (Jones 1975). Air pollution control in Pittsburgh had no effect until it rose to the county, then state level of jurisdiction. In this period, municipal ordinances on pollution were few and far between. Most covered only smoke pollution, not other industrial pollutants. Cities with smoke control ordinances, like Chicago, rarely enforced them (Davies 1969, 33). Figure 5.1 below shows the number of municipalities and counties with air pollution ordinances.

It became apparent that municipal governments were not equal to the task and, by the mid–1950s some state governments preempted the legislative authority of local governments. Thus the state governments favoured centralization at least to the level of state government, although not the federal government. As Figure 5.1 shows, the number of states with air pollution legislation began to increase dramatically after 1950. The sharp increase after 1965 is the result of federal legislation requiring that states have air pollution legislation on the books.

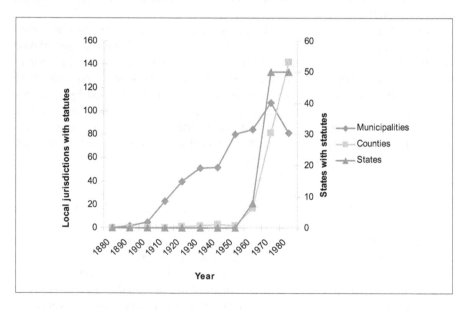

Figure 5.1 Air pollution control regulation in the United States
Source: Davies 1969, 126; Degler 1970; Stern 1982.

The Era of Local Action: Water Pollution

Water pollution fared little better under local control. Between 1850 and 1911, every major American city built sanitary sewers, in response to epidemics caused by contamination of wells. These discharged (usually untreated) wastewater to nearby bodies of water (Andrews 1999, 116). There were also some very early efforts at sewage treatment. By 1904, three per cent of the sewage of the American urban population received some treatment (US NRC 1939, 10). However, when effective water purification technology appeared, most cities made capital investments in this technology and neglected wastewater treatment, which was thought to be superfluous.[1] Some experts recognized the dangers of discharging sewage into sources of water supply: a 1908 report to the Conference of State and Provincial Boards of Health opposed this common practice (Andrews 1999, 124).

The construction of municipal sewage treatment plants was also inhibited by difficulties of local financing. Between 1932 and 1938, the percentage of the populated connected to sewage treatment almost doubled because of US federal grants/loans for public works projects. (See Figure 5.2 below.) The progress made in wastewater treatment in those six years exceeded that in the preceding 25 years (US NRC 1939, iii, 8). The federal government subsidies for the construction of municipal sewage treatment plants continued until mid–1990s when they were to be phased out under the *Clean Water Act* of 1987 (McConnell and Schwarz 1992, 55).

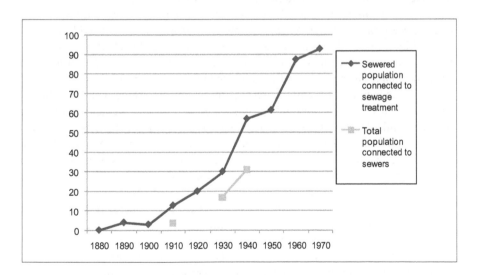

Figure 5.2 Per cent of US population connected to sewage treatment
Source: Tarr 1996, 174; US NRC 1939.

1   Oddly enough, a modern analysis of 122 American cities in 1910 shows that cities with sewage treatment had lower rates of mortality than those with drinking water filtration (Gaspari and Woolf 1985).

Action by State Governments: Air Pollution

State action managed to reduce some of the most egregious cases of pollution. Pennsylvania and California did witness improvements in air quality. Between 1940 and 1966, American and European cities experienced several severe pollution episodes which, taken together, resulted in thousands of excess deaths. In the worst cases, smoke and sulfur dioxide levels were so high that people were vomiting (Cassell 1968). Smog emergencies were very frightening and all signs indicated that things would get worse before they got better. Some progress was made in addressing air pollution emergencies, such as legislation to temporarily close factories. But these measures were not designed to reduce 'normal' pollution that people were exposed to on a daily basis, only to provide emergency measures to avert disaster.

Severe air pollution episodes were what first brought the US federal government into the air quality domain. The cause of such episodes was poorly understood. Very little was known about the health effects of pollution, beyond the immediate statistics about excess deaths in the wake of such emergencies. The United States Public Health Service was asked to initiate research into the causes and effects of pollution in the early 1950s. At this point, federal action was wholly limited to research. In theory, the states could have pooled their resources to fund research but there is no indication this was ever contemplated. Presumably because of the public goods nature of research and the very small number of scientists with any competence in this area, the federal government took the lead.

Among the first findings of this research was that automobiles were the single largest cause of Los Angeles's smog problems, by a very wide margin. Initial efforts centered on setting car emissions standards for California. At first, car manufacturers favoured state action over federal legislation. Manufacturers planned to make cleaner cars for the California market but not for the remainder of the American market. This attitude helped build support for federal action. During Senate hearings, Senator Edmund Muskie asked why cleaner vehicles should not be made available to all Americans. He found the industries' reply unsatisfactory.

Although Muskie strongly believed that pollution control should remain a state responsibility, Senate hearings persuaded him that an exception must be made for car emissions standards. When it became apparent that other states besides California might set car emissions standards, the auto industry changed its strategy. By 1965, the industry favoured preemptive federal regulation. The industry gambled that a single federal standard, even a strict one, was better than 50 different state standards (Dyck 1970).

The 1950s and 1960s saw unprecedented economic growth and rising incomes for all classes. Booming industries and mushrooming personal consumption helped to ensure that environmental problems continued to get worse. Air and water quality continued to deteriorate.

Very few states responded aggressively to this situation, California being the notable exception. California was spending 57 per cent of the *total* $2 million being spent in state programs, and 55 per cent of the total $8 million local expenditure was being spent by California localities (80 per cent of which was spent by Los Angeles) (US Congress. Senate. Committee on Public Works 1963, 31–37).

By 1960, only eight states had statewide air pollution laws (Degler 1979, 2). As of 1963, 15 states had air pollution control authorities, and only California and Oregon had extensive standards (Dyck 1970, 141). What legislation there was, did not address pollution from specific industries. Legislation did not set emissions standards, only prescriptions about the darkness of smoke permitted. There is little evidence to suggest that the legislation was widely enforced. We can infer this from the minuscule staffs and meager budgets of the state boards of health: '[b]y 1965, after two years experience with the *Clean Air Act*, 40 states had established some kind of air pollution program budgeting over $5,000 each, but only nine of those States had a regulatory program'(Jones 1975, 119).

Table 5.1 Water pollution status of American regions in 1937

Water pollution status of regions	Name of region	Number of basins in region
All basins in region seriously polluted	Lake Erie	1
	Hudson	1
	Central New England	4
	New Jersey–New York Coast	1
	Delaware	1
	Susquehanna	1
	Upper Ohio–Beaver	2
	Ohio–Northern Tributaries	3
	Wabash–Lower Ohio	2
	Kanawha	1
	Southern California Coastal	1
50% of basins in region seriously polluted, 50% somewhat polluted	Upper Great Lakes	3
	Hudson Bay (Red River of the North)	3
	Lower Arkansas River	3
	Chesapeake Bay	4
	Ohio–Southern Tributaries	3
	Upper Mississippi	7
50% basins in region seriously polluted, 50% not polluted	Lake Ontario/St. Lawrence	3

Source: US NRC 1939, 40–41.

Action by State Governments: Water Pollution

A series of reports by the United States National Resources Committee, a New Deal agency, describe the status of water resources in the US in the 1930s. The Committee found that, as of 1937, most watersheds north of Mason–Dixon line and east of the Mississippi were classified as seriously polluted. The most serious sources of water pollution were untreated sewage, leaching from mines and wastes from industry. The pattern of pollution corresponded to the greatest areas of urbanization, in part because many heavily polluting industries were located in urban areas. The report noted that '[t]here is one great belt in the northeastern United States in which pollution problems are outstanding and four outlying basins in which the problems are no less serious'(US NRC 1939, 40–41). Table 5.1 above lists the polluted basins, of 116 basins surveyed nationally.

The report also surveyed that status of water pollution control efforts at the state level. Only 15 of 48 states had an adequate legislative framework for regulating water pollution. In seven states, state agencies acted only in an advisory capacity. The report found deficiencies in the laws of 33 states.

One major deficiency was exemptions in state legislation for municipalities or for particular industries. In some cases, industries were exempt because no adequate treatment technology existed, but in other cases the exemption was simply 'in response to resistance by the industries' (US NRC 1939, 70). States also failed to grant sufficient authority to agencies responsible for enforcement. Many state laws did not have sufficiently clear standards of required water quality or permissible pollution to permit enforcement and judicial interpretation. Another significant obstacle to industrial pollution control was the lack of research on abatement technology, either by industry or state governments (US NRC 1939, 72).

It might be argued that state and local governments did not act because their action was being crowded out by federal initiatives and expenditures. To the contrary, the federal government was very reluctant to intervene. Environmental historian Joel Tarr, author of the definitive history of waste management in the US, finds little evidence that federal grants for sewage treatment crowded out investments that other levels of government would have made (Tarr 1996). This finding is supported by an econometric analysis of US wastewater treatment expenditures (McConnell and Schwarz 1992).

As late as 1960, President Eisenhower vetoed federal funding for sewage treatment, arguing that water pollution was a 'uniquely local' problem (Sundquist 1969). Every president from Franklin Delano Roosevelt to Richard Nixon went on record to assert the primacy of the states in pollution control. Truman and Eisenhower strongly believed that pollution control was a local responsibility. Eisenhower created a national commission to identify policy domains which could be returned to the states, to reverse the centralization which had occurred during the Great Depression and World War II. The commission identified sewage treatment plant construction as a purely local concern, with no appropriate role for the federal government. This recommended change was eventually ignored, and by 1982 sewage treatment plant construction grants were the federal government's most expensive public works program (Sylves 1982, 122).

The Federal Government's Increasing Role

Distilling the historical record, we find that, prior to federal intervention, state and local governments had a fairly poor record in pollution control. State legislation was very limited in scope and often larded through with exemptions. State agencies lacked the necessary authority, resources and staff to enforce the law. The states conducted no research into pollution control. Most importantly, ubiquitous concerns about competitiveness inhibited more aggressive action by state governments.

From the 1950s to the mid–1960s, the federal government became increasingly active in the fight against pollution, albeit in a strictly circumscribed role. Federal action was limited to research, support of interstate efforts and very substantial grants for sewage treatment construction. The 1963 *Clean Air Act* sought to encourage interstate compacts on air pollution, by providing large grants to any such agencies, up to 75 per cent of the cost. Examining air pollution policies of the era, Charles Jones noted that no interstate agencies developed as a result of the act and 'perhaps none should have been expected'(1975, 118).

Early federal legislation mandated the use of enforcement conferences in addressing interstate air and water pollution disputes (See Tables 5.4, 5.5 and 5.6 for examples). This was in keeping with the federal government's 'hands off' role as a facilitator with its interest limited to interstate pollution. Fifty–one conferences were held across the country between 1957 and 1971 (Zwick and Benstock 1971, 432). Such conferences were usually called by one of the states affected by the interstate pollution. The Enforcement Conferences on Interstate Pollution, held across the US on the status of specific rivers or lakes, exposed the feebleness of state and interstate action. The states individually and collectively were extremely reluctant to act on pollution, particularly if it originated from industry. The fact that these conferences, with their sometimes stomach churning testimony on pollution, were open to the public did much to raise awareness of the severity of pollution problems (Graham Jr. 1966). Having to answer the questions of federal officials in the presence of the general public and the media, greatly discomfited state and industry representatives and exposed their cosy relationship.

Although the federal government favoured state leadership, it became evident that the states would not act without considerable federal government prodding. In the 1967 sessions of state legislatures, 100 air pollution control bills were introduced (Jones 1975, 68). By 1970, over 200 state local and regional air pollution agencies had been established. However, as of 1970 *no state* had completed all of the requirements of the 1967 *Clean Air Act* for setting and enforcing of air quality standards (Jones 1975, 128). The Johnson administration's original version of the 1967 *Clean Air Act* had mandated national emissions standards, but Senator

Edmund Muskie rejected this. The final version of the Act still limited the federal role to research, funding and facilitation of state and interstate efforts.

Edmund Muskie, the leading Senator on pollution control issues, favoured state action over federal action. He saw federal action as the last resort, to be used only in the case of demonstrated state incapacity or unwillingness to act. By 1970, he reluctantly concluded that a stronger federal role was unavoidable. As a result, the 1970 *Clean Air Act* Amendments required the establishment of national air quality standards and national emissions standards for air pollution sources, similar to what had been initially been proposed in 1967. By 1974, the passage of the *Safe Drinking Water Act* meant that the federal government had mandated minimum standards in most areas of water and air pollution control (Zimmerman 1991, 95). Although Presidents and federal legislators had resisted setting minimum national standards, federal pre–emption of air and water pollution control coincides with significant improvements in pollution reduction. However, federal pre–emption was preceded by a long period of experimentation with interstate compacts, particularly in water pollution.

Interstate Compacts as a Critical Case

For many years, the American federal government sought to encourage interstate compacts in a variety of policy areas, including pollution control. What did compacts accomplish? What can we learn from the experience of interstate compacts? Compacts have been proposed for a range of problems and many have been implemented, with varying degrees of success. The nature of the problem strongly affects the chances of creating an effective compact. The most ill fated compacts were those designed to limit economic competition on taxes, wages or regulations.

Interstate compacts within the United States constitute a critical case for binding interjurisdictional cooperation: if interjurisdictional cooperation fails here, where could it succeed? The American federal system provides for a variety of enforcement mechanisms. The primary enforcement mechanism available to the states is the United States Supreme Court. Some compacts provide for enforcement in district federal courts or a court in any compact signatory state. Thus, because compacts can be enforced states should not be deterred from participation because of fear of defection.

In the US, interstate compacts have a long history. They were provided for both in the Articles of Confederation (1777) as well as the Constitution (1787). They reached the peak of their popularity in the 1920s and 1930s, when they were hailed as the solution to problems which were too big for individual states to address. Compacts would promote efficient administration and modern, scientific policies, without centralization of power in the federal government. For some of these problems, interstate compacts were an alternative to constitutional amendment and almost a necessity. Prior to the New Deal, the Constitution prohibited federal action in many policy areas, particularly regulatory policies. During this period, there was also strong support for states' rights and state level initiative, from some Progressives as well as most conservatives.

Frankfurter and Landis, envisioned an America governed, not by the states or the federal government alone, but through regional arrangements tailored to the specific circumstances of America's regions. Interstate compacts were at the heart of this vision (Frankfurter and Landis 1925, 34). This model advocated regional action on the basis of regional problems and regional attributes. We do not observe regional government of the type they envisaged because the most salient feature is the existing allocation of powers along political boundaries, not the physical boundaries of a particular problem. The jurisdictional allocation of powers determines how an interjurisdictional problem will be addressed, not vice versa. Furthermore, regional proximity is not sufficient to overcome the collective action problem of cooperation among state governments.

At least one of Frankfurter and Landis's contemporaries thought their faith in compacts was misplaced. Governor of Pennsylvania Gifford Pinchot was both a conservationist and a Progressive who opposed centralization. He made the following observations about interstate compacts for utility regulation:

> I used to be open to persuasion in the matter of compacts till I came to understand their capacity for interminable delays. Then I saw the error of my ways. The Boulder Dam Compact, dealing with one dam in one river, took something like a dozen years to put through. While I do not question the sincerity of those who proposed this solution, they should consider such facts (US NRC 1935, 46).

One reason for the 'interminable delays' is that interstate compacts, like international treaties, rely on unanimous decision–making. The first interstate water allocation compact on the Colorado River was only possible after the federal government allowed Arizona to be overridden, by setting a 6/7 voting rule for passage, instead of the usual unanimity requirement. Individual states will not enter into an agreement that they think might leave them worse off.

Interstate compacts were proposed for a wide variety of problems (see Table 5.2). It is instructive to observe which proposed compacts were successfully negotiated and employed. Compacts with redistributive implications or those involving economic competition were doomed to failure. Compacts or uniform laws dealing with commerce or business transactions were widely implemented. Compacts characterized by reciprocity also fared well. Environmental compacts will be addressed subsequently.

Types of Interstate Compacts: Failures and Successes

Compacts to limit economic competition   During the desperate years of the Great Depression, compacts were proposed to address labor and economic issues. Of these, only a compact to limit oil production (similar to a cartel) was successful, although it had to be modified once it was ruled unconstitutional. Compacts to guarantee a minimum wage, maximum hours and to regulate child labor failed, even though opinion polls of the time indicated that there was majority public support for such measures in all regions in the US (Tables 5.3a and 5.3b).

Table 5.2 Problems for which interstate compacts were suggested during the twentieth century

Issue	Compact formally proposed?	Compact concluded?
	Economic regulation	
Banking	Yes (Uniform bank chartering act)	No
Child labor law	Yes	Yes, but failed to be ratified by member states
Cotton production	No	No
Insurance		Nation–wide uniform laws
Maximum hours of work	Yes	Yes, but failed to be ratified by member states
Milk supply regulation	Yes	One is currently in force
Minimum wage laws	Yes	Yes, but failed to be ratified by member states
Occupational safety		No
Oil production (limitation of)	Yes	Yes, but ruled unconstitutional in 1935. Replaced by Interstate Compact to Conserve Oil and Gas
Prison labor (limits on use of)	Yes	Yes, part of *National Industrial Recovery Act* which was ruled unconstitutional by the Supreme Court
Public utilities regulation. Proposed by Governor Pinchot to create a national electrical grid.	Yes, but *Wheeler–Rayburn Act* of 1936 centralized authority on this issue.	No
Securities (sale of)		No
Social Security (residency requirements)	Yes	No
Standardization (electrical equipment)		No
Standardization (agricultural products)		No
Taxation	Yes	Yes, two
Tobacco production	Yes	Yes, but only ratified by one state
Trade barriers (Interstate)	Interstate conference held in 1939	No
Trucking regulation (weights and dimensions)	Yes	No. State truck size and weight limits were preempted by Congress in 1982

Unemployment insurance	Yes	No
Welfare (transient relief)	Yes	Yes, limited compact after the Depression

Commercial transactions

Accounting	Yes	Yes
Commercial transactions		Nationwide uniform laws (*Uniform Fiduciaries Act*)

Reciprocal compacts

Civil defense and disaster mutual aid	Yes	Yes
Crime prevention	Yes	No
Criminal law	Yes	Yes (hot pursuit, extradition, out–of–state witnesses)
Drivers' licenses	Yes	Yes
Drug abuse	Yes	Yes
Forest fire prevention	Yes	Yes
Higher education	Yes	Yes
Juvenile delinquents	Yes	Yes
Mental health (reciprocal privileges to state mental hospitals)	Yes	Yes
Supervision of parolees and probationers	Yes	Yes

Infrastructure compacts

Bridges	Yes	Yes
Highways		No
Navigation (control and improvement)	Yes	New York Port Authority
Parks	Yes	Yes

Natural resource management

Fisheries	Yes	Yes, several
Flood control	Yes	Yes, a couple
Irrigation		No
Natural resources conservation	Yes	Yes
Pest control (for example, boll weevil)		No
Soil erosion	No	No
Water supply	Yes	Yes, many in Western US

Environment compacts

Air pollution control	Yes	No
Nuclear energy regulation	Yes	No
Pollution of streams	Yes	Yes, several regional compacts
Radioactive waste (low–level)	Yes	Yes

Source: Council of State Governments 1956; Dodd 1936; Zimmerman and Wendell 1951; Zimmerman 1996.

Table 5.3a American opinion polls on federal authority to regulate labour

'Do you favor or oppose an amendment to the Constitution giving Congress the power to limit, regulate and prohibit the labour persons under 18?'

Date	Group surveyed	Favour	Oppose
20 May 1936	Residents of New York State (NY legislature did not ratify Child Labour Compact)	63%	37%
21 May 1936	Farmers	46%	54%
24 May 1936	Sample of US Population	61%	39%
21 February 1937	Sample of US Population	76%	24%

Source: Gallup Institute 1972, 23, 28, 50.

Table 5.3b American opinion polls on federal authority to regulate labour

'Do you favor an amendment to the Constitution giving Congress the power to regulate minimum wages?'

Date	Group surveyed	Breakdown of group	Favour	Oppose
19 July 1936	Sample of US Population		70%	30%
		New England	67%	33%
		Middle Atlantic	70%	30%
		East Central	69%	31%
		West Central	66%	34%
		South	69%	31%
		Mountain	70%	30%
		Pacific	74%	26%

Source: Gallup Institute 1972, 23, 28, 50.

Only a few states agreed to these compacts and subsequently both compacts failed to obtain ratification in state legislatures. The states negotiating the compacts were Northeastern and Midwestern states with high labor standards, which sought to limit the use of child labor in industrializing southern states. Although a few southern states attended the conferences, there was never any chance that they would sign, much less ratify the compact (US NRC 1935, 46).

Commercial transactions Compacts or uniform laws governing commercial transactions continue to be successful. These are private laws, based upon the common law, governing legal relationships among private individuals. Although state agencies do not administer such laws, state courts will provide remedies (Zimmerman 1996, 188). These compacts address problems where uniform action by all states is the preferred outcome. In 1896, the National Conference of Commissioners on Uniform State Laws, one of the first of many uniform law organizations, developed the *Negotiable Instruments Act* (Riechmann 1978). As early as 1925, all the states were signatories. The Uniform Commercial Code which succeeded the Act in 1951, had been enacted by 49 states by 1967 (Zimmerman 1996, 189). These particular compacts or agreements are like coordination games, with few distributive consequences.

Reciprocal compacts: Diffuse reciprocity Compacts which grant reciprocal privileges or reciprocal access to services also have a very high level of state participation. Examples of the former include the Drivers License Compacts and the Interstate Compact for Supervision of Parolees and Probationers, which was the first compact to be joined by all 50 states. Some service provision compacts are intended to create economies of scale for smaller or less populated states. For example, the Western Regional Education Compact allows students from its 15 member states to attend any of its three veterinary schools.

Environmental compacts and other similar compacts The following types of compacts have the greatest relevance to environmental issues: infrastructure, river allocation, flood control and water pollution control. Infrastructure compacts, where defection from the agreement would be very costly to other parties, seem to have been successful. Many compacts have been negotiated for river water allocation and these compacts remain relevant and in force. Flood control compacts, in contrast, were a disaster. About a dozen water pollution control compacts were negotiated and implemented. No air pollution control compacts were ever implemented. No comprehensive interstate environment compact was ever implemented.

Infrastructure compacts and the problem of defection There are many infrastructure compacts for tasks such as building and operating interstate toll bridges. These are examples of deep cooperation. Here defection could be very costly but there is little incentive for unilateral defection. If the partner state has built its half of the bridge, a state does not benefit by failing to build its half. (In game theory, these compacts represent assurance game.) Interestingly, these compacts look like ordinary contracts and have specific terms and conditions. These compact commissions are also granted far more power by state governments than those of environmental compacts: '[i]nterstate commissions ... are considered bodies 'corporate and politic' by express provisions in most interstate agreements and by judicial recognition of the United States Supreme Court'.[2] As such, they may be provided the authority to enter into contracts and to promulgate rules affecting their internal organization and functions' (Curlin 1972, 344). From the

---

2   *Petty v. Tennessee-Missouri Bridge Commission,* 359 US 275,277 (1959).

example of infrastructure compacts, it seems in cases where defection would be very costly, states create strong institutions which bind the participants.

Interstate Water Allocation Compacts:
Decision–making in the Shadow of Hierarchy

The largest single class of compacts involves water allocation. They are found predominantly in the western US, where water is scarce and irrigation is important, although more recent allocation compacts are in the South. States need secure water supplies for their citizens and farmers. For some states, an interstate water allocation compact may be preferable to a long and uncertain battle before the US Supreme Court. However, water allocation compacts can hardly be said to be a rapid response; they have taken an average of eight years to become effective (Zimmerman 1996, 36). Some Supreme Court decisions have mandated the negotiation of interstate water allocation compacts.

Water allocation compacts are more numerous and more successful than water pollution control compacts, even though such compacts are effectively zero–sum games, because two states can't both withdraw the same unit of water in the river. There are two reasons for the popularity and apparent effectiveness of these compacts. The first is their very narrow scope. Such compacts deal with a finite set of players and authoritatively define a single objective variable: the volume of water to be allocated to each. Once the watershed has been identified, the number of states involved is readily apparent. The compacts apportion water according to objectively quantifiable measures, such as a certain quantity of acre–feet of water per year.[3] These quantities are measured and a state will sue other states if it does not obtain its allotted quota of water. It is relatively easy to monitor compliance with the allocation.

The second reason for the popularity of river basin compacts is the unpalatability of the alternative: protracted and expensive litigation. In the absence of a negotiated outcome, the Supreme Court will authoritatively allocate water supplies, based on legal precedent. Decision–making in the shadow of legal hierarchy means that there is less benefit in refusing to participate in a compact. It is usually cheaper and less risky for states to reach agreement among themselves than to leave their fate in the hands of the judicial system. In the western US, water is allocated by the doctrine of prior use and this decision rule can be disadvantageous to states that developed later.

---

3   Problems have arisen when the estimates of total water have been based on years of unusually high rainfall.

Flood Control Compacts:
Failure to Reach Agreement in the Presence of Mutual Gains

The states proved unable to address flood control through interstate compacts. The federal *Flood Control Act* of 1936 re–asserted state primacy in this area and encouraged the formation of interstate compacts while providing federal funds. (Projects authorized by this act were qualified to receive federal funds for construction and maintenance, payment of damages, and purchase of rights of way, which can be very costly.) During this era, some states were incapable of carrying out flood control projects lying entirely within their own state. Individual states (or localities) refused to agree to the permanent flooding which was a natural consequence of the construction of dams and reservoirs. The only successful instance of local cooperation on flood control was the Miami River floodplain near Dayton, Ohio (Platt 1980). Without the exercise of central authority, disputes over dam siting often proved impossible to resolve, apparently the 1930s equivalent of NIMBY (Not In My Backyard).

These irresolvable disputes persisted even where flood control was an urgent priority. Over a 20 year period, most of the New England states were ravaged by a series of floods. Even after three severe floods with substantial property damage and loss of life, the affected states were unable to reach agreement on effective flood control measures (Leuchtenburg 1953). Cost was not the obstacle because federal grants were available for dam construction and maintenance.

Nor was it simply an upstream/downstream conflict: upstream communities also suffered great losses during some of the floods. Despite extensive damage in Vermont, the government of Vermont consistently refused to allow the construction of flood control dams. Each locality identified by the US Army Corps of Engineers as a possible dam site refused to allow a dam to be built. These dams would be designed and constructed by the US Army Corps of Engineers (US ACE), once a state agreed on sites.

One reason why the New England states were incapable of cooperation on dam construction was their opposition in principle to federal intervention. They viewed federal funding and the involvement of the US ACE as unacceptable federal intrusion. New England state governments were vehemently opposed to federal action in general and the Tennessee Valley Authority (TVA) in particular, a powerful and autonomous federal agency responsible for developing hydroelectric power and the economy in general in the Tennessee Valley. There was widespread fear that a TVA–type New Deal agency would be created for New England.

Many opinion leaders and politicians in New England were remarkably parochial and short–sighted: a Boston paper stated in 1937: 'Floods, no matter how disastrous, are more welcome than the New Deal control of New England [electrical] power'(Leuchtenburg 1953, 71). The Governor of Vermont recommended that, in lieu of permanently flooding farmland in beautiful Vermont, federal money would be better spent moving whole cities, such as Springfield MA, to higher ground (Leuchtenburg 1953, 179). During this era, many New Englanders advocated states' rights as strongly as Southerners did during the days of slavery (Patterson 1969).

In the case of flood control, benefits which could be achieved *only* through cooperation were not sufficient to generate cooperation through interstate compacts. This was true even though the federal government was ready to provide ample technical and financial assistance. Leuchtenburg concludes: '[t]he essential feature of the interstate compact is that it attempts to resolve conflicts by giving one of the parties to that conflict a veto. Unless the other states will agree to the terms of any particular state, that state can refuse to enter into an agreement' (1953, 251). Vermont's consent would only have been possible with huge side payments (perhaps not even then), in addition to the financial assistance already provided by the federal government. Unlike the case of pollution control, there is no way that inadequate flood control could confer an economic benefit on a state, at the expense of other states. The issue was ultimately resolved when flood control came under federal jurisdiction, removing the need for state unanimity.

Interstate Environment Compacts

The limitations of the interstate compact approach to environmental protection can be seen in two ways. The first is the limited coverage of such compacts. The only environmental protection compacts to come into existence were those addressing water pollution and the disposal of low–level radioactive waste. This chapter will not examine the Low–level Radioactive Waste Compacts because they did not evolve bottom up from state action, but represent delegation from the center. Federal legislation assigned each state membership in one of ten interstate compacts which were intended to identify new sites for disposal of low–level radioactive waste (a classic NIMBY type of problem). These compacts are generally considered a failure. Since the passage of the federal legislation in 1980, no new facilities have opened (see Rabe 1990, 130; Zimmermann 1996).

Of all the rivers or lakes affected by interstate water pollution, only half were ever covered by interstate compacts (see Table 5.6 for large, polluted interstate rivers not covered by compacts). The second shortcoming is the limited effectiveness of the compacts which were implemented. Only a handful were ever effective in reducing water pollution (see Table 5.4) Most compacts included poorly defined pollution control objectives or control measures totally lacking in enforcement power (see Table 5.5). The lack of controls on industrial pollution is particularly conspicuous in these compacts.

Although water pollution control compacts are the only ones to have been implemented, two other kinds of compacts were proposed. Because water pollution compacts had shown some promise, the federal government wished to promote similar institutions for air pollution control. No air pollution compact had been proposed until the federal government offered to fund 75 per cent of the cost of such an agency, a measure introduced in the 1963 *Clean Air Act*. Only five interstate air compacts were ever formally proposed. None received federal approval.

The second type of environmental compact was an overarching compact to cover all aspects of environmental policy in all 50 states, the Interstate Environment Compact (1970). Congress refused to grant approval to this compact because it was completely hollow, simply exhorting states to cooperate on the issue.[4] It did not commit the states to do anything nor did it constrain them from any course of action. It was widely seen as a cynical tactic to halt the federal environmental juggernaut, which was picking up speed at this time.

On the whole, the efforts of the states individually and collectively in compacts, were unable to prevent the continuing deterioration of water quality. The significant improvement in water quality, observed since the 1970s, is almost entirely attributable to federal legislation (Adler 1993; Patrick 1992). Very few states had water quality standards as stringent as those introduced by federal legislation. In any event, the environmentally activist states, such as California, were in no way representative of the states as a whole. The average state took little action on pollution, regardless of the severity of its pollution problems. Interstate cooperation was primarily effective in reducing water pollution from untreated sewage and here the task was greatly eased by generous federal funding for sewage treatment plant construction.

Water Pollution Control Compacts

There have been approximately thirteen compacts which have addressed water pollution control. Water pollution control is the primary objective of eight of these compacts. Of these eight, only three had a measurable effect on pollution reduction (see Table 5.4). The most effective compacts are the Delaware River Compact, ORSANCO (Ohio River) and the Interstate Sanitation Compact (New York Harbor). These compacts grant more powers to the compact commission than ineffective compacts. The five which are less effective generally grant no powers to the compact commission (see Table 5.5). The single exception is the Susquehanna, which grants approximately the same powers as the effective ones, although it also asserts that the states, not the compact, have primacy. These compacts show that standard–setting and enforcement authority are necessary but not sufficient conditions for effective pollution control.

Water pollution control compacts share some common characteristics. Most require unanimity in decision–making except for the Delaware and Susquehanna Compacts, which use majority voting and three–quarter majority respectively (US GAO 1981).

Two of the three most effective compacts (the Delaware and ORSANCO) owe their origins to serious threats to drinking water for major cities. During drought

---

4   US Congress. Senate. Committee on Public Works. Subcommittee on Air and Water Pollution 1972.

conditions, there was insufficient stream flow to dilute the pollution, seriously compromising drinking water quality in the cities downstream. These compacts were preceded by at least one Supreme Court case between the signatories to the compact. These compacts also cover heavily industrialized watersheds. New York State is a member of all three effective compacts, but also two ineffective compacts: the Susquehanna and Great Lakes compacts.

Even the effective compacts, such as ORSANCO, were hesitant to address industrial pollution. Before federal preemption, compact agencies sought to reduce pollution from industrial sources by relying on persuasion and voluntary measures by industry. Where interstate compacts have the power to enforce regulations, these powers have been used as a last resort. Although enforcement actions were rarely taken, when they were used, it was usually against municipalities with inadequate sewage treatment and not industrial polluters. The greatest success of water pollution compacts has been the reduction of municipal pollution, with the help of substantial grants in aid from the federal government for wastewater treatment plant construction. States and cities are intensely sensitive to the possibility of losing industrial investment to other jurisdictions, even through indirect means such as bond issues to finance water treatment.

Cooperative efforts for pollution control on lakes   The experience of cooperative pollution control on lakes is very disappointing. It could be argued that the difficulties in negotiating water pollution control agreements arise from the intrinsic conflict of interest between upstream and downstream parties in the river system. However, interstate water bodies not subject to upstream/downstream divisions fare no better. There are no agreements covering lakes which are as ambitious or as effective as ORSANCO. Interstate compacts governing Lake Tahoe and the Great Lakes are both weak. The presence of symmetrical interests was not sufficient to engender effective action.

Lake Tahoe, on the California–Nevada border, is an interesting case because the purity of the lake was an important factor in making Lake Tahoe a tourist destination. The unique beauty of this ancient lake was threatened by increasing sewage contamination, as was the drinking water supply, drawn from the lake. The interstate compact proved almost powerless to prevent the further degradation of the lake (Strong 1984). The rapidly increasing deterioration was slowed by the passage of the 1972 *Water Pollution Control Act* Amendments, which set national sewage treatment standards. The federal government's generous grant program for sewage treatment construction also helped. The failure of this compact is particularly sobering because only two states were involved and tourism, not heavy industry, forms the economic base of the area.

In fairness to the Lake Tahoe interstate compact, its greatest weakness was that the Governing Council represented local interests above state interests, and these were the interests that were most biased in favour of development (Baxter 1973). Attempts to reduce sewage pollution depended on the cooperation of the five counties around the lake. A revised compact which reduced the dominance of local interests performed somewhat better in safeguarding Lake Tahoe's environment. However even this modest step was prompted by threats from President Carter that the federal government would step in if California and Nevada were unable to safeguard this national treasure (Strong 1984, 192). Even at a local level where there was great potential for mutual gain, fears over economic competition blocked cooperation.

The Great Lakes, involving more states and much more industrial pollution, represent an even bleaker scenario. The Great Lakes Compact was particularly ineffective because the role of the compact commission was purely recommendatory.[5] Although Canada borders several of the Great Lakes, the Canadian sources contribute less to the total pollution than American sources do (Kehoe 1997). The pollution problems of Lake Michigan, which lies wholly within US territory, were no less intractable than those of the other Great Lakes. Most of the improvement witnessed in the Great Lakes over the last 30 years, such as the recovery of Lake Erie, is attributable to federal action (Adler 1993).

The role of the federal government in interstate compacts and cooperation   The federal government supported interstate compacts over federal action, from the 1920s (Scarpino 1985). Conditions for interstate pollution compacts became even more favourable in the 1950s and 1960s when compact agencies were eligible for substantial grants from the federal government, at a higher proportion of funding than grants to the states. In addition to providing enforcement capacity, the federal government explicitly encouraged compacts with the transfer of many millions of dollars. Even under these favourable conditions few compacts were negotiated and the environmental performance of these varied widely.

Of the few pollution control compacts negotiated, some were motivated by the possibility of increased federal involvement. States proved able to cooperate only when they faced the imminent threat of federal preemption. On several occasions, states negotiated weak interstate compacts hoping to stave off federal action. The New England Interstate Water Pollution Control Compact (1947) and the Interstate Environment Compact (1970) are the principal examples. These compacts are either recommendatory or completely hollow, not requiring the states to do anything they would not have done in the absence of the compact. In these cases, the states had shown themselves incapable of acting individually or collectively, over many decades in some cases. These compacts showed that the only thing the states could agree on was that they did not want federal action.

Interstate environmental compacts: Too soon to judge?   It might be argued that federal pre–emption occurred too soon for pollution control compacts to prove their worth. On the contrary, interstate pollution control compacts had decades in which to prove their efficacy for highly polluted river basins in the eastern US. The

---

5   This compact is distinct from that concluded in 2008, which focuses on preventing diversion of water out of the Great Lakes watershed.

US Public Health Service sought to facilitate informal interstate cooperation on 'stream pollution' as early as 1916. The streams and lakes governed by compacts, and others not covered by compacts, were identified as interstate pollution problems by 1935 (US NRC 1935).

Interstate water pollution compacts did not address stream pollution in a comprehensive manner. Enormous polluted watersheds, sources of drinking water for millions, such as the Mississippi and Missouri Rivers, were not covered at all (see Table 5.6). Of the compacts which were successfully negotiated, a minority of those had any measurable effect on pollution reduction (see Tables 5.4 and 5.5). The most effective compacts granted powers to the compact commission. In general, compacts were constructed in such a way that they were not binding on the states in the compact or did not provide for enforcement of environmental protection within those states.

More recent attempts at interstate cooperation: OTAG   The states have had more recent opportunities to prove they can improve on the federal government's record. Interstate pollution problems persist: much of the Northeast's smog originates in Midwestern coal–fired utilities. There is a strong upstream–downstream dynamic to this pattern, akin to pollution on a river. Emissions reductions in Midwestern states would reduce smog in the Northeast. However, these upstream states, which may not have any smog problems of their own, have no incentive to reduce their emissions in order to solely benefit those downstream. The Northeastern states could impose more stringent limits on local sources of these pollutants, but local sources would consider that unfair when those sources are not the primary causes of the problem.

In principle, nothing prevents the states from addressing this problem through compact mechanisms: the *Clean Air Act* amendments assign the states the leading role in implementing the amendments. In 1995, the Ozone Transport Assessment Group (OTAG), was created, representing all 37 states east of the Mississippi. It was a self–conscious effort to create a process for dealing with multi–state environmental problems, without involving the direct involvement of the federal government (Keating and Farrell 1999, iii). The goal was to arrive at a solution, through consensus, which would be more economically efficient than uniform national emissions standards, for example an emissions trading scheme for smog causing pollutants.

OTAG did extensive modeling and deliberated on the issue for several years. It represented one of the largest exercises in air pollution data gathering, modeling and analysis ever conducted: over 400 scenarios were examined (Keating and Farrell 1999, 1). The process involved extensive stakeholder consultations. However, a schism emerged between the Northeastern states on the one hand, and Midwestern states with coal–fired utilities and southeastern coal producing states on the other hand (Jones and Bucher 1998, 11–6).

In its analysis, OTAG was unwilling to examine the possibility of geographically differentiated controls. The group only studied control strategies which would have applied uniformly to the whole OTAG region (Keating and Farrell 1999, 127). Economists generally consider such uniform standards economically inefficient. The Northeastern states and those in the Lake Michigan area were unwilling to discuss regionally differentiated controls. One of the leaders of the OTAG process was reluctant to press the issue, for fear that the OTAG group would be unable to reach any consensus, no matter how weak (Keating and Farrell 1999, 127). In its choice of modeling studies, the group was also unwilling to examine the appropriate balance between region–wide controls and local controls.

In June 1997, OTAG released its scientific conclusions and recommendations to the Environmental Protection Agency. OTAG did not present any conclusions about the relative importance of long range transport to local smog problems, even though a quantitative analysis of this issue was one of the main objectives of the process. Many of the recommendations were very broad and relatively weak. By a margin of 32 to seven, the states voted in favour of emissions controls, but without giving any details about how stringent those emissions controls should be. OTAG recommended that the EPA set stringent national product standards for fuels, vehicles and products such as coatings. The group did not make recommendations for actions to be undertaken by the states, individually or collectively. In the report, the states did not commit to reducing emissions. A 1999 assessment of the OTAG process concluded that the interstate process generated the lowest common denominator outcome, explicitly comparing the outcome to the kind of broad, nonbinding proposals of the sort commonly seen in international treaties (Keating and Farrell 1999, 124).

There was no agreement on an interstate emissions trading scheme. This is no surprise. We do not see such strong measures because there is no win–win outcome possible, without large side payments from the Northeast to the Midwest, Pennsylvania and West Virginia. That would be an example of a Coasean bargain where the victims of pollution pay the polluters. Such payments have not been suggested because they would be politically unacceptable in the Northeastern states.

In August 1999, Republican governors in the Northeast announced a compromise solution in the OTAG impasse (Pérez-Peña 1999a, A1). However, within a week the compromise solution had evaporated and state officials were in agreement that the issue would be resolved in the courts. The Director of Michigan's Department of Environmental Quality said '[t]here won't be any settlement because there isn't any middle ground'(Pérez-Peña 1999b, A16).

The outcome of the OTAG experience is consistent with past evidence on attempts at cooperative solutions. As noted in Chapter 4, directional externalities are rarely addressed cooperatively (Barrett 1991, 166–7). Economically efficient solutions, such as regionally differentiated standards, never received serious consideration because they were considered to be politically unacceptable. Only uniform solutions were contemplated, probably for reasons of perceived fairness.

The States in Contemporary American Environmental Policy

Those arguing for noncentralized environmental policy may concede the weakness of state policies in the past while arguing that things have changed. What do decentralization advocates claim has changed? State governments now have the institutional capacity to make environmental policy. Without federal interference, state policies would better reflect public preferences. States would be better able to tailor policies. Advocates of greater state autonomy point to states that set standards in excess of federal minimums to argue that decentralization would not cause a race to the bottom. While it is difficult to make counterfactual arguments about what state policy *might* be like, without federal oversight, there are signs in the current system that indicate that decentralization would lead to lower levels of environmental protection. The federal government still plays a valuable role in environmental protection.

Public Opinion on the Environment

Unfortunately, there were almost no national public opinion polls on the environment prior to 1965 (Jones 1975, 141). In the US, systematic public opinion polling had begun in 1935. There are still few systematic polls about the environment on a state by state basis (Ringquist 1993, 109). Thus it is not possible to determine whether state or federal environmental policies are more representative.

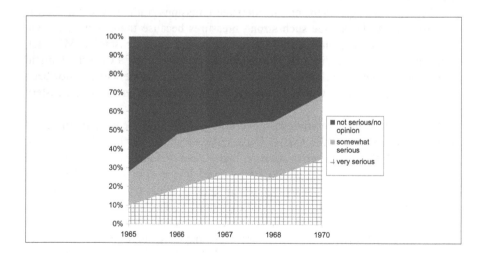

Figure 5.3 'Compared to other parts of the country (US), how serious do you think the problem of air pollution is in this area?'
Source: Davies 1975, 82.

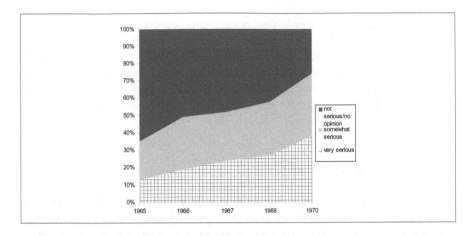

**Figure 5.4 'Compared to other parts of the country (US), how serious
      do you think the problem of water pollution is in this area?'**
Source: Davies 1975, 82.

From earliest polling data, it is clear that, by 1970, there was a pervasive sense of crisis about the state of the environment. (see Figures 5.3 and 5.4). In June 1970, the percentage of the population that considered air pollution to be 'very or somewhat serious' in their area had surged from 28 per cent in 1965, to 69 per cent. The corresponding figures for water pollution were 35 per cent in 1965 and 74 per cent in 1970 (Davies 1975, 82). On the basis of this polling data, we observe that growing concern about the environment occurred at the same time as increasing federal preemption.

State Environmental Policy and Adaptation to Local Conditions

Some states have set standards more stringent than federal minimum standards, however this does not prove that they would have the same standards in the *absence* of federal floor standards. The argument requires a counterfactual: we cannot infer what states would do in the absence of the federal government from evidence *in the presence* of the federal government. These higher standards are set in the shadow of a hierarchy. It is possible that such standards will only exist when there is guarantee that standards in other states will not fall below a certain minimum. There are also states, such as North Carolina, whose legislation *prohibits* state standards more stringent than the federal minimum (Lowry 1992, 77). It seems plausible, that given the opportunity, these states would introduce laxer standards or perhaps eliminate standards entirely. The removal of federal minimum standards could spur a race to the bottom, at least by some states.

William Lowry examined the conditions under which state governments exceed the minimum federal standards. He examined four areas of pollution control policy: stationary source air pollution, point source water pollution, mobile source air pollution and non–point source water pollution. Lowry found that some states did act, but only under particular conditions: '[t]he lower the level of interstate competition in a policy area, the more likely [it is] that leading state programs supersede federal guidelines'(Lowry 1992, 126). He concluded that concerns about capital flight and attracting new investment are very important in explaining the pattern of state leadership.

The Role of Federal Oversight

American federal government mandates appear to guarantee a higher level of environmental protection and also greater transparency and public access. Although the federal government delegates administration of most environmental policies to the states, it retains a powerful oversight function. The importance of this function can be seen in the implementation of the *Safe Drinking Water Act.*

If the federal government's role should be limited to addressing transboundary pollution, the setting drinking water standards lies outside that ambit. However, the US federal government is thought to have a legitimate interest in standard setting for drinking water because of the public health issue. Interestingly, one of the European Commission's most invasive environmental directives governs drinking water quality from the tap. Canada has national (voluntary) guidelines for drinking water quality but no national regulations.

Drinking Water

The US *Safe Drinking Water Act*, dating from 1974, is intended to ensure minimum standards of drinking water and to provide Americans with the information to assess the performance of their local water supplier. Prior to the passage of the Act, the General Accounting Office (US GAO) conducted a review of state programs for drinking water in six states. The report found that potentially dangerous water was being delivered to some customers, particularly by those systems serving 5,000 customers or less. Eighty–one of 446 of water supply systems (18 per cent) exceeded federal bacteria standards for more than two months of the year. Another nine per cent exceeded bacteria limits for one month a year. The US GAO also judged the water quality monitoring activities in five of the six states to be insufficient, mostly due to insufficient numbers of staff. Only 23 of 67 interstate water carriers had been inspected by the states during 1972, even though the EPA recommended annual inspections. Over 85 per cent of systems did not meet federal bacteriological or sampling requirements, a finding which applied to large and small water supply systems.[6]

---

6   US Congress. Senate. Committee on Public Works. 1975, 136.

In 1999, an EPA audit showed that state authorities failed to report 88 per cent of the infractions of drinking water standards which they were required to report to the federal government's Safe Drinking Water Information System (SDWIS) (Eisler 1999a, 1A). SDWIS is the basis of the federal government's efforts to disseminate information on local water quality to citizens. It is also the basis for assessing the performance of America's water supply system as a whole and to enable learning across jurisdictions.

The states argued that the violations not reported were mostly missed deadlines. They also argued that, if authorities were able to resolve the problem with the water supplier responsible for the violation, there was no need to report the violation to the public. However, among the infractions not reported to the SDWIS were:

- 30 per cent of all cases where water systems exceeded legal limits for fecal coliform bacteria;
- 80 per cent of violations of nonbacterial contamination, such as pesticide residues;
- 90 per cent of all cases where water systems failed to properly treat or filter their supplies (Eisler 1999b, 5A).

Thus although state governments are closer to the people, in this case they were more likely to conceal information from local customers and tax payers. The implicit rule seems to be that what people don't know, won't hurt them. In the absence of federal requirements, much less data on local water quality would be available. These findings suggest that the federal government still plays a valuable role in protecting public drinking water supplies.

Conclusion

State governments were neither particularly active nor effective in addressing pollution in the first half of the twentieth century. While we can document the existence of pollution problems and water borne diseases, we cannot determine the extent to which state and local policies accurately reflected public preferences in this period. The data simply do not exist. As public concern about pollution intensified, the federal government's role continued to grow, culminating in the setting of minimum national standards for many types of air and water pollution. Federal action, through standard setting, research and ample subsidies for construction of sewage treatment plants, resulted in significant improvements in environmental quality.

The fact that a federal role was necessary in the past does not necessarily mean a similar role is necessary today. The literature suggests that the federal government is not yet superfluous. Federal minimum standards may stimulate the tailoring of state policies to local conditions. State level policies continue to be very sensitive to concerns about competitiveness: federal floor standards assuage that concern enough to permit some states to exceed federal standards. Federal oversight of drinking water contributes to public safety and greater transparency. While this federal role is unnecessary according an externalities model of pollution control, it appears that citizens would have far less information about their local water supply and less protection in the absence of federal involvement.

Although a handful of interstate compacts achieved some reduction in pollution, the US experience with compacts is not a ringing endorsement of interjurisdictional cooperation. This is true even when the number of parties was small, mutual gains were possible, interests were symmetrical, financial support was available and contracts could be enforced. Some failures of cooperation can be attributed to parochialism and shortsightedness, such as the Lake Tahoe compact or the New England flood control compacts. Most seem to have foundered on the shoals of concerns about economic competitiveness. Although a compact might insure that states in a watershed did not take investment from one another because of disparate pollution control requirements, the compact could not prevent investment from going to states *outside of the compact*. The role of environmental regulation in business location decisions is disputed and is considered by many scholars to be minimal. However, these concerns are perennially on politicians' minds. Loss of investment and jobs are always cited as reasons against requiring pollution control.

Table 5.4 Effective water pollution control compacts

Name of stream or lake (Name of compact, if different)	Surrounding states	Date compact proposed	Federal enforcement conference	Compact ratified (Year of congressional approval)	Members of compact	Powers of compact	Effectiveness of compact
Delaware River Basin Compact	DE NJ NY PA	1925, 1927, 1936, 1959	None	1961	DE NJ NY PA Feds	Can set standards and enforce them.[1] Broad powers and no unit veto.[2] Compact commission may institute judicial action in its own name to compel compliance.[3] Has authority to construct sewage treatment works.[4]	Primary goal was to cap the amount of water New York City could draw from the river. The more water NYC took, the more polluted the river was as it flowed out of NY, where other large cities drew their drinking water from it. Droughts made matters worse. Supreme Court cases in 1929 and in 1954.
Ohio River (ORSANCO)	IN IL NY OH WV KY PA VA TN AL NC	1936	None	1948 (1940)	IN IL NY OH WV KY PA VA Feds	Can set standards and can enforce them but each state has a veto. Enforcement powers rarely used, persuasion preferred.	Primarily effective in reducing the amount of untreated sewage dumped into the river. Less effect on industrial pollution, particularly acid mine drainage.
New York Harbor (Interstate Sanitation Compact)	NY NJ CT	1935	Raritan Bay (1961, 1962)	1941 (1935)	NY NJ CT	Commission has power to compel compliance with standards it sets but prefers persuasion. Each state has a veto.	Primarily effective in upgrading sewage treatment facilities. But as late 1966, 22 per cent of the waste discharged received no treatment.[5]

Notes: [1] Muys 1971, 52; [2] Muys 1971, 58; [3] Curlin 1972, 352; [4] Davies 1970, 137; [5] Chambers 1969, 54.

Table 5.5 Less effective water pollution control compacts

Name of stream or lake (Name of Compact, if different)	Surrounding states	Proposed compact	Federal enforcement conference	Compact ratified (Year of congressional approval)	Members of compact	Powers of compact	Effectiveness of compact
Susquehanna River Basin	NY PA MD		None	1970	NY PA MD Feds	Modelled on Delaware Compact but primary responsibility for the control of water pollution is left to the individual states.[1]	Focused on water allocation, maintenance of flow and some monitoring.
(New England Interstate Water Poll'n Control Compact)	MA CT VT ME NH RI	1935	None	1947	MA CT VT ME NH RI	Classifies interstate waters according to classifications submitted by the states. Has no authority to compel a state to classify its waters for a particular use or to enforce the standard set.	'Has been a flat failure, as any objective look at the polluted streams of that region will disclose'.[2] Very slow; after 20 years the states had not agreed on classifications for eleven streams, including part of the Connecticut River.[3]
(Potomac Valley Conservancy District Compact)	DC MD PA VA WV	by 1939	None	1940, 1970	DC MD PA VA WV Feds	Makes recommendations which must be approved by two of three commissioners from the affected state.[4]	Does water quality monitoring.
Lake Tahoe (Tahoe Regional Planning Compact)	CA NV	1955, 1967 (Lake Tahoe Interstate Water Conference Committee formed 1931)	None	1969, 1980	CA NV	Regional planning which would help reduce flow of sewage into Lake Tahoe.[5] Each state can veto.[6]	First Compact which gave a lot of power to the five local governments had no effect on limiting water pollution. National sewage treatment standards have had greatest impact.[7]

Compact	States	Year		Ratification	Year	Authority	Impact
Great Lakes Basin Compact	IL, MI, IN, WI, MN, PA, NY, OH	1955	Lake Erie 1965, Lake Michigan 1968	1955 – IL, MI, IN, WI, MN 1956 – PA 1960 – NY 1963 – OH	1968	No standard-setting authority.[8] No power to compel action by a state.[9]	No evidence of any impact on pollution reduction.[10]
(Tennessee River Basin Water Pollution Control Compact)	TN KY MS AL GA NC VA		None	1955-TN 1957- KY MS. no approval from AL GA NC VA	1958	Commission can establish water quality standards for various classifications of use.[11]	Nil. Never became active due to insufficient ratifications.
(Klamath River Basin Compact)	CA OR		None	CA OR. President appoints one non-voting member[12]	1957	Water allocation primary goal. Decisions made by consensus.	No evidence of action on water pollution.
(Arkansas River Basin Compact)	CO KS	1965	None	CO KS	1966	Water allocation primary goal. To recommend and coordinate. No standard-setting authority.[13]	No evidence of action on water pollution.[14]
Bear River	ID UT WY	1955	None	ID UT WY	1958	Water allocation primary goal.	No evidence of action on water pollution.
Red River of the North	ND, SD, MN		None	ND, SD, MN	1938	No standard-setting authority.[15]	No evidence of action on pollution.[16]

Notes: [1] Chambers 1969, 53; [2] Graham 1966, 218; [3] Chambers 1969, 64; [4] Chambers 1969, 61; [5] Strong 1984, 130; [6] Baxter 1974, 7; [7] Strong 1984, 135; [8] Curlin 1972, 348; [9] Chambers 1969, 64; [10] Kehoe 1997, 28; [11] Curlin 1972, 349; [12] Chambers 1969, 60; [13] Curlin 1972, 347; [14] Curlin 1972, 354; [15] Curlin 1972, 347; [16] Curlin 1972, 354.

Table 5.6 Polluted interstate streams which could have formed the basis of a compact

Name of stream or lake	Surrounding states	Date when pollution problem was first recognized	Was it the object of an informal water quality agreement?	A major source of drinking water?	Was it the object of a federal interstate pollution enforcement conference?	Interstate compact proposed?	Did compact receive federal approval?
Mississippi River (Upper)	MN WI IA IL MO	1906 – the development that created the Twin Cities had already turned the River into an open sewer.[1]	Yes. Upper Mississippi Drainage Basin Sanitation Agreement (1935) MN IA WI IL MO	Yes – Minneapolis, St. Louis	No	No	
Mississippi River (Lower)	MO IL KY TN AR MS LA	Early twentieth century	No	Yes	Yes	No	
Missouri River	MT ND SD NE IA KS MO (Basin includes WY CO, Alberta, Saskatchewan)	1910 – sewage pollution identified as primary source of typhoid outbreaks in the region.[2]	No	Yes, but judged a potential hazard by EPA in 1971.[3]	Yes	1953. Revised draft prepared by Council of State Governments.[4] Basin development compact. SD KS MO MT NE ND (core), Feds, potentially IA CO MN	Never reached that stage

Notes: [1] Scarpino 1985, 163; [2] Schmulbach et al. 1992, 146; [3] Schmulbach et al. 1992, 147; [4] McKinley 1955, 351.

Chapter 6

Switzerland: The Power of Referenda in a Noncentralized System

If asked to describe Switzerland, most people envision snow–capped mountains and pristine alpine villages and meadows. Switzerland has the image of being an oasis away from pollution. The reality is quite different: one book on the subject was entitled *Das Märchen von der sauberen Schweiz* [The Fairytale of a Clean Switzerland) (Brandenburger et al 1982). Switzerland has had environmental problems for some time. In comparison with Germany or France, however, Switzerland was late in effectively addressed its pollution problems. How could this happen in a country with many strong environmental groups? The explanation lies in Switzerland's strong institutional bias towards noncentralization. Switzerland's noncentralized tendency exacerbated environmental collective action problems, which were ultimately overcome by referenda that amended the Swiss Constitution.

Why look at Switzerland? It is a small country, anomalous even within Europe. It is this very uniqueness that makes it of interest to political scientists. First, it represents one of the least centralized federations in the world (Filippov et al 2004). Sovereignty resides with the cantons: they have far more autonomy than most subnational governments. Switzerland has a smaller national government than most OECD nations, in terms of state expenditures as a fraction of GDP. Second, it has used referenda longer than any other country and still uses them frequently. In keeping with Switzerland's anti–majoritarian traditions, a constitutional referendum requires a double majority for passage. The proposition must be approved by a simple majority of the whole population and in a majority of the cantons. Thus any measure approved by a referendum enjoys broad public support, not just the support of a simple majority.

Switzerland is an example of increasingly centralized environmental policymaking over time, a trend which also occurred in the United States. However, in the American case, this centralization might be seen as secular trend of centralization in all areas of public policy and not a deliberate choice about environmental policy. This explanation is not plausible in the Swiss case. Switzerland has real constitutional and political barriers to centralization. On the whole, the Swiss have 'been very reluctant to extend competencies to the federal state'(Armingeon 2000, 118). Environmental policy represents an exception to this trend, where centralization has occurred through a series of referenda. This demonstrates substantial and widespread support for both stronger environmental policies and a stronger federal role in environmental policymaking.

The chapter begins by describing the environmental and cultural factors that lead us to expect noncentralized but effective environmental policies in Switzerland. The second section discusses Switzerland's anti–centralist tendencies, and the consequences this noncentralization has had for the country's water quality. The third section presents what should have been an easy case for local pollution control: fluoride pollution in a single canton, Valais. The fourth section contrasts the potential for intercantonal cooperation on the environment with the historical record, showing almost no cooperation. The fifth section then surveys Swiss referenda on the environment over the last century to demonstrate that the Swiss have repeatedly sought to improve environmental quality and transfer jurisdiction on the environment to the federal level. The final section summarizes the implications of the case for the larger argument on noncentralization and environmental protection.

While Switzerland is home to some of Europe's most spectacular natural scenery, it has also suffered significant environmental stresses. It has high population density, with 178 people per km². This is less than Germany or the UK, but more than France or Denmark. As Switzerland industrialized in the late nineteenth century, its pollution problems began. Already in 1875, the Swiss Confederation demanded that the cantons study the toxicity of municipal wastewater (Walter 1990, 107). An 1883 Swiss study of the Rhine found that much of the effluent entering the river was toxic to fish (Walter 1990, 107). By 1897, Zürich had suffered its first severe algal bloom, due to eutrophication of Lake Zürich (Walter 1990, 180). Air pollution problems appeared somewhat later. In 1914, farmers brought suit against an aluminum smelter in the canton Valais for damage to crops from air pollution (Walter 1990, 183).

If any country should be able to carry out effective, noncentralized environmental policy, it should be Switzerland. Switzerland has long been a wealthy country. Its economy has never been based on pollution intensive activities: there are relatively few heavy industries such as steel making. Since the nineteenth century, the picturesque landscape has been the mainstay of an important tourist industry.

In addition to an economic base conducive to environmental protection, there is evidence of strong popular support for environmental protection. As of 1990, 11 per cent of the population belonged to an environmental organization, the third highest participation rate among the 48 countries in the World Values Survey[1] (Inglehart et al. 1997, V26). Switzerland's oldest nature conservation groups started in 1905 and 1909 (Walter 1990, 99). Despite a long tradition of environmental groups, from the 1950s to the 1980s, Switzerland's pollution problems continued to worsen. Between 1950 and 1965, sulphur dioxide emissions increased two–and–a–half–fold, declining sharply only after 1980. From 1950, NO_x emissions

---

1   Only the Netherlands and Denmark had higher rates of participation.

increased seven–fold, peaking in 1980.[2] In contrast, the US and Germany had halted or reversed deterioration in these pollutants by the 1970s.

Swiss Public Support for Environmental Protection

Although the Swiss have had large environmental groups for decades – a good proxy measure for the level of popular environmental awareness and concern – it is difficult to measure Swiss public opinion on the environment, especially over time.[3] Data is scarce because Swiss opinion polling is a fairly recent phenomenon: there was little opinion polling on environmental issues prior to 1986.

The few polls that exist suggest strong popular support for environmental issues in the early 1970s and again in the mid–1980s when acid rain and forest damage became salient (Schenkel 1998, 112). Polls from 1972 show that 85 per cent of German speaking Swiss thought that the environment was the most important problem facing Switzerland. In the same year, 79 per cent of those polled thought that water pollution urgently needed to be cleaned up; 65 per cent thought air pollution urgently needed to be cleaned up (Zürcher 1978, 89). Table 6.1 below shows a high level of concern about the environment, with a majority continuing to favour devoting more government resources to the environment. It should be noted that Switzerland had few economic problems in the 1980s but entered a recession in the early 1990s that might explain the decline in the environment's rank as the most severe problem.

Table 6.1 Relative importance of environmental problems

		1986	1987	1988	1990	1991	1992	1993
The environment is the most [severe] problem	Percentage who agree	39%	49%	54%	29%	27%	15%	9%
	Overall ranking of environment as a problem	1st	1st	1st	1st	1st	3rd	3rd
Other state tasks should be cut in favour of the environment	Percentage who agree	79%	81%	75%	79%	76%	70%	59%

Source: Univox (sample 600–700 persons per year) (Schenkel 1998, 113).

---

2   Switzerland. BUWAL (Federal Office of Environment, Forests and Landscape) 1994, 240.

3   For example, Switzerland participated in only one of the six questions about the environment in the 1990 World Values Survey (Inglehart et al. 1998).

Another indication of popular support for environmental protection is popular initiatives on the environment. Popular initiatives are referenda triggered by public petition, subject to a sufficient number of signatures. In general, Swiss social movements focus their political mobilization on popular initiatives, rather than lobbying politicians or other forms of activism (Armingeon 2000, 114). Environment, energy, animal rights and traffic restrictions were the subject of 32.2 per cent of the popular initiatives submitted between 1974 and 1993 (Kobach 1994, 144). Only a minority of these popular initiatives have passed but the relatively high proportion of popular initiatives devoted to the environment indicates demonstrates that the environmental movement has the size and capacity to put forward these initiatives and to get them on the ballot. With one exception, these popular initiatives sought to strengthen environmental protection and many proposed fairly extreme measures, such as the decommissioning of all nuclear power plants.

From Fragmentation to Centralization:
The Impact on Swiss Water Pollution

Because the Swiss show high levels of concern about the environment, but generally oppose centralization, we would expect that if anyone could make noncentralized environmental policy work, the Swiss could. Most areas of Swiss policy are characterized by fragmentation. Environmental policy is no exception. The Swiss political system is designed around multiple elite bargaining forums that are intended to prevent cleavage along linguistic and regional lines (Armingeon 1998, 170). The strength of sub–national governments (cantons) means that the federal budget is substantially smaller than in other OECD countries, although total government spending as a percentage of GDP is not.

Swiss water policy highlights the impact of fragmentation. Even when federal legislation has been enacted, it has been undermined by inconsistent implementation at the cantonal level. There is wide variation between the cantons, but even the performance of the most activist cantons leaves something to be desired. To overcome inconsistent implementation, Swiss water policy has become progressively more centralized over many decades. Greater federal intervention has been accompanied by measurable improvements in environmental quality. Eutrophication peaked in the 1970s and by the late 1990s, most lakes and rivers showed good water quality (Mauch and Reynard 2002, 3).

Early water protection laws were ineffective, particularly in the absence of regulations or federal enforcement. Switzerland's first water protection measures were introduced to protect fish: fish dieoffs from water pollution were documented as early as the 1890s. Zürich and another canton created water police in the 1880s (Walter 1990, 104). In 1889, a federal law on fisheries mandated sand filtration of industrial emissions. There is no evidence this law was ever implemented by the cantons: many lacked water police to enforce the law. The law was supplemented by a regulation in 1925. However, this regulation allowed cantons to exempt firms from treatment requirements if such requirements would harm the interests of the firm. The federal government was not authorized to punish non–compliance. The few firms caught polluting preferred paying the fine to building waste treatment facilities (Walter 1990, 225).

Pollution was caused not only by unrestricted industrial wastes, but also untreated municipal sewage. In general, municipal wastewater requires more extensive, higher level sewage treatment in Switzerland than in North America because of geography. Many Swiss cities are located on lakes, which are sometimes a source of drinking water. These lakes are very susceptible to eutrophication, thus most contemporary Swiss sewage treatment plants use tertiary or quaternary treatment to reduce phosphorus in effluent from the sewage treatment plants. Secondary treatment is considered sufficient in most places in the US and the European Union and is more common than tertiary treatment. Despite the greater vulnerability of the Swiss water resources, when addressing the issue at the local level, the Swiss were no more proactive on municipal sources of water pollution than they were on industrial sources.

Rates of Sewage Treatment in Switzerland

Sewage treatment facilities were rare in Switzerland and it was not until the 1980s that most of the population was connected to a sewage treatment plant (STP). Although any urban area will require some form of sewage treatment, the construction of sewage treatment plants was very uneven across the cantons. Until the late 1950s, the majority of STPs were in the canton of Zürich. The city of Zürich built its first STP in 1926. By 1950, there were 30 treatment plants in the country, over half of which were in the canton of Zürich. There was a clear need for sewage treatment plants, however, because by 1953, almost a dozen large Swiss lakes were deemed to be endangered by pollution. Nevertheless, the number of people served by STPs remained low. As late as 1963, several people in Zermatt contracted typhus from drinking water contaminated with human waste (Zürcher 1978, 31).

As shown in Figure 6.1 below, as water policy became more centralized, rates of sewage treatment increased dramatically. Given the vulnerability of its lakes and its population density, there is little to suggest that Switzerland's 12 per cent rate of sewage treatment in 1964 was adequate. By contrast, West Germany already had 50 per cent of its population connected to STP by 1965. The steep increase in Swiss STPs coincided with the introduction of federal subsidies for the construction of sewage treatment plants in 1962 (EAWAG 1981, 37). The subsidy mechanism was made more effective after in 1972. These subsidies were subsequently phased out in an effort to implement the Polluter Pays Principle. Under the old funding formula, the federal government paid 15–45 per cent of costs (30 per cent on average) and the cantons paid up to 50 per cent (35 per cent on average), with the municipality paying the remainder (EAWAG 1981, 42). By the early 1990s, the percentage of the population served by a STP was very high, given that many rural dwellers will never be connected to municipal sewers.

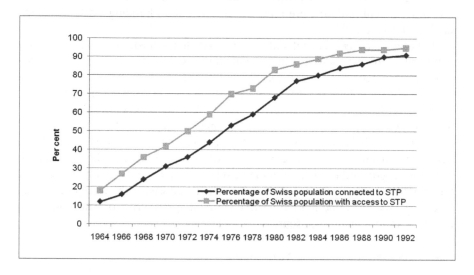

Figure 6.1 Municipal wastewater treatment in Switzerland
Source: BUWAL 1994.

What were the politics behind changes in Swiss water policy? Groups concerned about water pollution began to appear in the 1930s. Groups for water pollution control favoured a stronger federal role, whereas industry groups wanted the cantons to retain as much authority as possible. In 1933, a group for the protection of Lake Zürich called for a constitutional amendment, giving the federal government jurisdiction over water pollution (Walter 1990, 194). In 1949, a national group was formed to raise public awareness about water pollution problems (Bussman 1981, 156). A draft constitutional amendment and draft federal water pollution law were circulated in the late 1940s.

The leading organization representing Swiss business (Schweizerische Handels und Industrie Verein) opposed federal legislation. They took the position that cantonal legislation could be effective and must be tried first. Even those who were not opposed to federal legislation on principle thought that federal legislation should just be a framework law (Bussman 1981, 159–60). This was in keeping with the Swiss antipathy towards centralization.

However, by 1955, federal water legislation became a reality. A 1953 referendum approved a constitutional amendment extending federal jurisdiction over water protection policy. The vote was four to one in favour and there was a majority in favour in every single canton. This result far exceeded the minimum requirement for an overall majority and a majority of cantons, demonstrating strong and universal support for the measure.

The amendment was put into effect by a 1955 federal law. It forbade the dumping of solid waste into water and provided for criminal penalties in the case of intentional pollution. The emission of industrial wastewater was subject to cantonal authorization. However, by 1965, it was apparent that water cleanup efforts from the 1955 bill could not keep pace with increasing pollution, leading to net deterioration in water quality. The Swiss Bar Association, Schweizerische Juristentag, concluded that the situation was due to defects in the 1955 law (Zürcher 1978, 34). In 1966, the federal government passed a regulation setting water quality standards for receiving waters. These were the first national standards set, although as always, they were subject to implementation by the cantons. There was no legal impediment to cantons taking regulatory action but, in general, they did not.

A small minority of cantons took action on water pollution prior to federal legislative requirements. As noted above, until the 1960s, sewage treatment was virtually nonexistent outside the canton of Zürich, even in cities. Few cantons took the initiative to address other sources of water pollution. Industry was not subject to water pollution controls.

In more recent times, water pollution from agriculture became a serious problem. We see a repeat of the experience with industrial and municipal sources of pollution: the cantons showed little initiative in addressing water pollution from agriculture. It appears that few cantons wanted to take on the farmers. As rates of industrial and municipal wastewater treatment have improved, the proportion of water pollution caused by agriculture has increased dramatically. Switzerland has very highly subsidized agriculture and farmers have incentives to farm intensively, both in livestock and crop production. More recently, runoff, particularly containing animal manure or fertilizer, has been the largest source of lake eutrophication.

Failure of Noncentralized Environmental Policymaking:
Air Pollution in the Canton of Valais

The experience with air pollution is consistent with developments in water pollution control. Switzerland should be conducive to noncentralized environmental policymaking. In theory, it should be possible for the cantons to develop environmental policies, appropriately tailored to their circumstances. Those making the case for subsidiarity or decentralization argue that local authorities better reflect local preferences and have better knowledge of local conditions. Thus, measures introduced by lower levels of government should be more appropriate than those introduced by national authorities.

The opposing view is that lower levels of government lack economies of scale in technical expertise and are more vulnerable to capture and fears of capital flight. In theory, local authorities should be able to arrive at appropriate regulations for an industry whose pollution is largely local in its impact and which is relatively immobile, thus rendering threats of exit less credible.

The case of fluoride pollution in the canton of Valais – home to all of Switzerland's aluminum smelting facilities – supports the claim that the cantons did *not* tailor environmental policy. This should, however, have been an *easy* case for local action. The pollution from smelting was known to be highly toxic but its effects were largely local, within the boundaries of the canton. The industry in question is not highly mobile, which should reduce fears of capital flight. There was no regulatory competition with other cantons because all the Swiss smelters were in the same canton. Yet, the aluminum industry of Valais was permitted to emit highly toxic pollutants with impunity, for over 60 years. In this case, there is also evidence of regulatory capture by industry, if not outright corruption, as officials easily switched from government jobs to positions in the companies and vice versa (Gerbely 1989, 251).

For over 60 years, the cantonal authorities in Valais turned a blind eye to severe fluoride pollution caused by aluminum smelters. They did not enforce their own cantonal laws nor the federal laws they were supposed to enforce. Although the Swiss smelters and the Swiss industry are small by world standards, one of the parent companies, Alusuisse, was one of the world's major aluminum producers for much of the twentieth century. The industry chose to locate in this canton because, at the turn of the twentieth century, it was developing its hydroelectric resources.[4] From an environmental point of view, however, Valais was not an ideal location for this type of facility. All the smelters lay in the Rhone river valley, resulting in little wind dispersion of air pollution. The toxic fluoride gases and particulates which were emitted, tended to remain in the valley.

Environmental damage appeared shortly after aluminum smelting began. The first facility, in Chippis, opened in 1905. By 1912, scientists had begun to investigate sickness in cattle in that area (Gerbely 1989, 248). The first of many court cases was brought against an aluminum smelter in 1914, by farmers whose crops had been damaged (Gasche 1981, 27). The case went all the way to federal court but it never reached judgment. The factory settled out of court – the preferred method for dealing with complaints. The factories would buy pieces of land or would offer compensation, in a single lump sum payment. Those receiving compensation agreed to keep the terms of their settlement secret and to relinquish the right to seek damages in court for past or future damages (Gasche 1981, 10–11).

By 1918, Swiss federal scientists substantiated that recently observed damage to forests in the area was caused by aluminum smelter emissions. In the same year, municipal officials and citizens lodged complaints about damage with one of the factories (Gerbely 1989, 248). However, cantonal authorities in the capital Sion hesitated, and did not dare take action against the factory, even though Sion had

---

4   Electricity is the most import input in aluminum smelting, hence smelters are often located where there is cheap, abundant hydroelectric power, such as Iceland, Norway and the province of Quebec.

also suffered damage. A 1917 report to the Valais government said '[s]ince our last report, the damage has extended and worsened, whole forests have had to be cut because they cannot withstand the gas emissions of the factory. Energetic measures must be taken if we are to prevent this idyllic landscape from being transformed into a wasteland'(Gerbely 1989, 249). By 1920, scientists had identified gaseous fluorine compounds as the cause of the damage (Gerbely 1989, 247).

Figure 6.2 below shows that this was only the beginning. From the time of the first complaints, aluminum production increased fivefold in Valais.

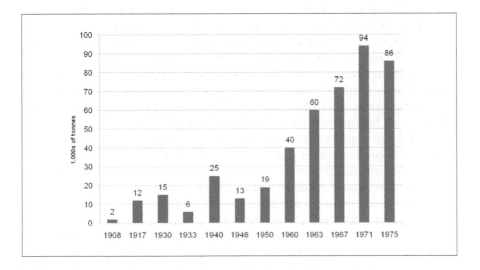

Figure 6.2 Annual raw aluminum production in Valais
Source: Gasche 1981.

By 1924, the press had taken sides against the factory. Caving in to public pressure, the canton passed a law, directing factories to 'prevent the release of substances which harm people, animals or the surrounding landscape'(translation mine) (Walter 1990, 184). On its face, 'no harm' is a very demanding criterion, which few pollution regulations aspire to meet. The Valais law forbade toxic emissions, without specifying any technical or economic criteria. The law was never applied against aluminum factories in Valais. Until superseded in the 1970s, the law remained a dead letter (Gerbely 1989, 249).

The damage caused by the smelters was not limited to cows, crops and forests. In 1935, the human occupational illness fluorosis was first documented. Of the 400 cases described in the international medical literature, 61 (15 per cent) were from Valais – a high proportion given the small scale of Switzerland's aluminum industry (Gasche 1981, 29). Because the smelting pots were open, located in large halls, workers were exposed to all the hazardous gases that accumulated. Better emissions control would protect not only the environment, but also the health of workers. The unions at the aluminum factories never lodged formal complaints, probably for fear of losing their jobs (Gasche 1981, 139–42).

In August 1953, farmers' frustration erupted into violence. Fruit and vegetable growers in the region rioted. Apricot farmers took over a train station and set freight cars on fire, necessitating the diversion of the Paris–Milan train. Federal troops were called in (Gerbely 1989, 235). In 1963, a different group of Swiss farmers protested by bringing their sick cows to Zürich. The Zürich headquarters of the aluminum firm in question got police protection. Farmers from Fricktal, not in Valais, sought compensation for damages from fluoride emissions from a Swiss owned smelter, located across the river in Rheinfelden, Germany.

The claims of the Fricktal farmers were treated quite differently from those in Valais, probably because the source of their damage lay across the border in Germany (even though the facility was Swiss owned). The farmers in Valais had to meet extremely restrictive standards to receive compensation for their sick cows: the Valais cantonal veterinarian had to diagnose fluorosis, which was subject to standards of proof much higher than those required in Fricktal. Relatively few farmers in Valais received compensation for their cattle. In constrast, the same firm (Alusuisse) had to buy 1,580 cows over a ten year period in Fricktal (Indermaur 1989, 55). It was not only the cantonal veterinarian who was under the influence of the industry. The Valais cantonal forester and director of the Valais cantonal agricultural college also appeared to be in Alusuisse's pocket (Gerbely 1989, 250).

In the 1960s, the three Valais smelters continued to expand production and pollution increased further. Farmers in Martigny were promised that, despite the expansion there, fluoride emissions from the factory would decline from 52.7 to 11.4 kg per day. Subsequent testing found emissions were six times that level (Gasche 1981, 19). In 1968, the nuns whose convent sat on a hill directly opposite the Chippis factory brought complaints, requesting compensation for headaches, dizziness and vertigo. The Alusuisse factory sent the Mother Superior 250 Swiss Francs. The nuns were subsequently found to have fluoride levels in their urine twice that of the general population. In 1975, Alusuisse gave the convent 30,000 Swiss Francs and pledged to give them 8,000 SF annually. The nuns withdrew their complaints (Gerbely 1989, 252).

Environmental conditions deteriorated during the 1970s. The farmers were spurred on to organize. By 1975, the smelters in Valais emitted more than a metric tonne (1,200 kg) of gaseous fluorine compounds per day. At times, the ambient levels of fluorides in the air in Valais were 30 times higher than standards would permit. The haze of pollution sometimes hung throughout most of the Rhone valley (Bauer 1989). Five years earlier, in 1970, after several crop failures, farmers in the Martigny area founded the League against Poisonous Factory Emissions (*Schutzverband gegen giftige Fabrikabgase*). The League spent its first five years writing letters to the factory and the authorities, to no avail (Gerbely 1989, 249). By 1975, the farmers threatened violence and municipal officials warned of the threat of riots.

The year 1975 marked a significant turning point. The municipality gave the League 100,000 SF and it hired an agricultural expert, Gerard Vuffray, who turned the league into an American–style citizens' movement. In 1975, the League lodged a complaint with the cantonal authorities. Cantonal official Arthur Bender replied that the group had no legal authority to intervene in the case (Brandenburger et al. 1982, 137). He implied that nothing could be done until the federal *Environmental Protection Act* passed (which did not happen until 1983).

However, the very same Arthur Bender had, in 1965, threatened the smaller Martigny factory with legal action on the basis of the federal Factory Law and cantonal legal authority over public health. At that time, Bender also noted that the cantonal law of 1924 was still valid. Bender later claimed a 1961 public health law had invalidated the 1924 law. But in 1976, when a politician demanded a cantonal *Environmental Protection Act*, Bender and others protested that the 1924 law was still in effect (Gasche 1981, 66). When pressed in an interview, a cantonal politician stated that the canton had done everything possible to implement the 1924 law (Brandenburger et al. 1982, 137).

In addition to being a capable organizer, Vuffray was a good researcher. He visited aluminum smelters in Europe, North America and South Africa. In 1977, he reported on fluoride pollution from the Valais smelters and the state of their pollution control measures. The report was a harsh indictment of the industry and the cantonal authorities. The League demanded that the industry adopt cleaner technology (closed smelting pots) and install dry scrubbers on its smokestacks. Facilities in Valais mostly had open pots and used wet scrubbers, which are useless against gaseous emissions. Vuffray showed how the factories in Chippis and Martigny had misled the authorities with false statements and data. Alusuisse had told federal and cantonal authorities that its filters retained 95 per cent of the effluent gases (Gasche 1981, 18). The data Vuffray collected showed this figure to be 50 per cent, at most.

In responding to the League's demands for better pollution control equipment, the Alusuisse directors replied this was impossible. They said that closed pot/dry scrubber technology was in the experimental stages. In 1976, Alusuisse claimed the Chippis factory was the most modern in Europe (Gasche 1981, 18). In fact, it was one of the worst in the world. By 1976, one–third of world production was made in closed pots. In several countries, such as the US, closed pots were required by law. These held back no less than 98 per cent of fluorine compounds. Vuffray's report noted that Alusuisse used this technology in its *American* factories, but not those in its native Switzerland (Gerbely 1989, 254).

Confronted with the evidence, Alusuisse pleaded poverty. Alusuisse claimed that in Chippis alone, retrofitting would cost 100 million SF and that Chippis had been operating at a loss for years. In fact, in the previous decade, Alusuisse had made 427 million SF in profits on its Swiss operations (Gasche 1981, 21). Citing a US Department of Commerce report, Vuffray countered that retrofitting all three facilities would cost at most 30 million francs. Predictably, Alusuisse claimed that meeting League's demands would lead to job losses and factory closure. A director of Alussuise claimed that 3,200 jobs were at risk in the town of Chippis, which had the oldest smelters. He had to retract this statement when it was revealed that even if there were job losses in Chippis, only the smelter portion of the operation, employing 150 of the factory's 3,000 workers, would be affected (Gerbely 1989, 257).

In January 1977, the federal labour agency embarrassed the cantonal government. The Bundesamt für Industrie, Gewerbe und Arbeit (BIGA), announced that the Valais authorities had not paid any attention to whether the factories upheld the 1966 federal labour law. This law protects not only factory workers, but also protects the factory's surroundings from damaging effects. The canton was responsible for enforcing this law.

BIGA asserted that the cantonal authorities had thwarted BIGA's orders and attempted to conceal facts from the federal authorities (Brandenburger et al. 1982, 140). In reply, the cantonal authorities claimed they had done everything possible, short of continuous monitoring (Gerbely 1989, 256). The cantonal authorities had not conducted any monitoring or inspection prior to mid–1970s. Previously, in 1976, the Valais environment protection office had admitted they knew nothing about the state of abatement equipment in the aluminum smelters, because the factories had not notified them about any breakdowns. They were only concerned about breakdowns, not regular operating conditions. Furthermore, until 1976, factory directors were notified prior to inspections (Gasche 1981, 70).

The League gave the canton a deadline: do something by March 1977. In the meantime, several commissions were at work on the issue. On 12 December 1977, the cantonal executive presented its conclusions. New smelting pots must have permits. New ovens could not be started while trees were in bloom. The Chippis factory must fix broken windows and leaky doors within three months (Gasche 1981, 71). The Alusuisse factory in Chippis did not make these changes by the deadline; they began work two months later in February 1978 (Gasche 1981, 22). In addition, the canton announced that the factories must ensure that their abatement technology functioned optimally and that fluoride emissions were minimized. This was a very vague order, which differed little from the proscriptions of the ineffective 1924 law (Gerbely 1989, 257).

The League was not impressed or satisfied. The Union of Valais Farmers reserved the right to resort to special action (that is, violence) if legal means failed to produce sufficient results (Gerbely 1989, 257). On 16 April 1978, a hydroelectric transmission tower was blown up. There were farmers' demonstrations and protests. In a meeting with a member of the cantonal executive, a farmer manhandled a cantonal official, tearing off his shirt collar (Gerbely 1989, 258).

That year, a federally appointed commission on fluoride pollution delivered its report. The report was given to the cantonal executive of Valais in April 1978, which decided to keep it secret. Six months later, activists obtained the report and publicized it (Gasche 1981, 23). The commission's findings largely corresponded to those of League (Gerbely 1989, 258): the Valais smelters were retaining barely half of their fluoride emissions. The report concluded that it was technically possible to reduce emissions to between one–quarter and one–seventh of existing emissions (Gasche 1981, 18). The report recommended that emissions be reduced to one–fifth of existing levels (Gerbely 1989, 258).

In October 1978, the Valais authorities mandated a cleanup of the three smelters (Gerbely 1989, 258). The two newer facilities had until 1981 to cut their emissions by 80 per cent. They were required to reduce total emissions from 619 to 129 tonnes fluoride per annum within three years (Gerbely 1989, 22). The oldest, most obsolete facility, Chippis, was given until 1993 to clean up – an additional 15 years. Chippis could emit fluoride at twice the rate of the other two factories until 1989.

Table 6.2 Emissions reduction requirements in Valais

Aluminum Factory	1975 emissions (kg fluoride/t aluminum)	New limit (kg fluoride/t aluminum)	Deadline
Steg	5.5	1.5	end of 1981
Martigny	7	1.5	end of 1981
Chippis	10	3.9	end of 1980
"	3.9	2.9	end of 1990
"	2.9	1.5	1994

Source: Gerbely 1989.

By the late–1970s, Alusuisse factories stepped up their air pollution abatement. The author of a critical history of the Alusuisse corporation has noted that the air pollution '... was not a technical or legal problem, rather solely and entirely a political one. First after citizen activism created political pressure, sufficient to outweigh the influence of Alusuisse, did the Wallis (Valais) cantonal government act' (translation mine) (Gerbely 1989, 237). On 2 March 1980 Alusuisse announced plans to invest 250 million SF in its facilities in Steg and Chippis. It also increased the capacity of its rolling mill in Siders by 40 per cent. There were no job losses (Gerbely 1989, 258).

If lower levels of government really are best suited to addressing pollution problems, then the situation in Valais should never have arisen.[5] Fluoride pollution there should have been easy for the cantonal authorities to address. First, it has long been known (since the 1910s) that fluoride pollution is highly toxic to plants, animals, forests and workers. Second, pollution control in this case would primarily benefit local residents, because emissions of this type are not usually transported long distances, especially in a mountain valley where there is little dispersion by wind. Third, although the smelters were the single largest private employer in

---

5   As noted above, federal efforts to address the problem were hamstrung by the requirement that the canton implement federal law, as well as the obstructionism of cantonal officials.

the canton of Valais, the risk of relocation was fairly low. Of the most important costs in aluminum production, only two – the costs of labour and electricity – vary significantly from one location to another.[6] The most important factor in locating a primary aluminum smelter is the cost of electricity, usually hydroelectricity, where Switzerland had an advantage over most of this period. Fourth, the facilities in Valais did not face a competitive *disadvantage* because of environmental control measures: these facilities were among *the last* in the OECD to impose effective emissions control technology.

This should be an easy case for local action: highly toxic pollution which is largely local in its effects, produced by a sector which is not mobile and is highly regulated elsewhere. However, the cantonal authorities were exceptionally slow to respond. The Chippis smelter opened in 1905 and damage had been identified by 1912. Yet it would be another 80 years (1994) before emissions from Chippis were subject to modern emissions controls. This response is certainly not rapid and was responsive primarily to industry, not local environmental conditions nor the views of local farmers.

The Record of Intercantonal Environmental Cooperation

While the cantons have been ineffective individually at achieving environmental protection, the collective efforts of cantons working together have been equally disappointing. Switzerland has the capacity for formal and informal cooperation between cantons. Formal cooperation can occur by means of a *Konkordat*, a treaty between two or more cantons. Most of Switzerland's major rivers run through several cantons. Several major lakes lie between them. Yet there are no formal agreements governing these intercantonal water resources. When referenda have demonstrated strong support for environmental protection across the diverse groups and various cantons, why have cooperative solutions to environmental problems failed to emerge? Why, for example, have there been almost no environmental *Konkordat* negotiated among cantons?

Like the American constitution, the Swiss constitution allows cantons to enter into treaties with one another. These instruments are growing in importance. The first of these was negotiated in 1803 (Frenkel 1986, 326). By 1980, 311 such agreements had been negotiated. By 2006, the number was estimated to exceed 760, with 10 per cent of these having been concluded since 1996 (Bochsler and Sciarini 2006, 29). Seventy–five per cent are agreements between two adjacent cantons; some encompass several cantons within a region. Only two dozen encompass 20 or more cantons (Bochsler and Sciarini 2006, 29).

---

6   This finding was confirmed by a survey of 145 aluminum smelters, comprising 93 per cent of world primary aluminum production. The findings of the survey done by AME Mineral Economics are summarized in 'Aluminum Smelter Costs', *Mining Journal*, 17 March 2000, 209.

Many resemble American interstate compacts: there are Konkordats for cooperation on physical infrastructure, universities, lotteries, professional accreditation, and so on (Vouga 1964). Intercantonal cooperation is seen as an important tool to prevent greater centralization, since 'continuously unresolved inter-cantonal problems provoke national initiatives justified by the failure of the cantons to deal with problems on the sub-national level' (Bolleyer 2006a, 14). Failed Konkordats have, in fact, been precursors to federal action (Bussmann 1986, 70.

However, when intercantonal cooperation on pollution was needed most, it did not occur. Relatively few of the Konkordat concern environmental issues, other than sewage treatment plant construction. During the period when Switzerland's water quality was deteriorating, no Konkordat were concluded on this issue. At the same time, the cantons negotiated Konkordats on other types of infrastructure investment: 24 such agreements were concluded between 1841 and 1951. The first infrastructure agreement for sewage treatment took place in 1957, when the federal government had already asserted jurisdiction in this area and was providing financial incentives. Between 1957 and 1988, twenty Konkordats on sewage treatment plant construction and operation were concluded, the majority after 1977.[7]

No Konkordat were created to set regional standards for water quality or emissions. The first Konkordat for common measures on water protection was concluded in 1974, between the cantons of Basel–city and Basel–country. In 1985, five cantons concluded an agreement for common measures for the protection of the lake Vierwaldstättersee. Thus intercantonal cooperation did not fill the void left by the individual cantons. Prior to federal legislation formal cooperation between cantons played almost no role in environmental protection.

Informal cooperation was not much more effective. Since 1897, Switzerland has had intercantonal working groups in a number of policy areas, which number more than 500 (Bochsler and Sciarini 2006, 27). Some include all cantons; others include all cantons within a region. The participants are department heads from cantonal governments; federal government counterparts are not members, although they are usually invited to the meetings. Of these working groups, the most important for the environment is the Swiss Construction, Planning and Environment Directors' Conference (BPUK) (Frenkel 1986, 330). These working groups are generally very informal; few have a secretariat. (In 2006, the secretariat of the BPUK had a full time staff of one (Bochsler and Sciarini 2006, 27). These groups are focused on information exchange about their common challenges (Bochsler and Sciarini 2006, 24). The resolutions from the working groups are not binding (Frenkel 1986, 332).

On water pollution, information cooperation at the intercantonal level had some impact but this cooperation was made effective by the looming threat of a larger federal role. A scholar who studied this cooperation called it 'centralization

---

7   University of Fribourg database of Konkordaten. Thanks to Daniel Bochsler for the data. http://federalism.unifr.ch/concordat/ge/index.html.

from below': the cooperative intercantonal water pollution working group *wanted* a strong federal role (Bussmann 1981). After the 1953 constitutional amendment extended federal authority to water pollution control, there was a period of many years without effective federal water pollution legislation. Technocrats at the cantonal level cooperated to develop minimum standards of sewage treatment and minimum national effluent standards. These standards were explicitly developed with the hope that they would be incorporated into federal legislation (which they were) (Bussmann 1981, 231). The officials at the cantonal level did not want this to remain a matter of cantonal cooperation.

The cantonal water pollution experts cooperated to develop minimum sewage treatment standards that were eventually adopted in federal regulations. They were unable to achieve consensus on water quality objectives, only water treatment objectives: they were able to agree on standards pertaining to effort but not environmental outcomes. They agreed that secondary sewage treatment facilities should be constructed in all population centers. For working group participants, it was important to know that other participants had to meet the same treatment standard, even if this represented a waste of money in some regions and an underinvestment in other regions with bigger pollution problems. While they could reach agreement on a uniform technical standard, this group proved unable to develop standards tailored to local environmental conditions. This undermines one of the major arguments against national standard setting: subnational actors should be better able to tailor measures to local environmental conditions.

It appears that a standard of secondary sewage treatment represented a focal point in bargaining. A level playing field with regard to treatment standards was the only outcome that could be reached by consensus. It is also worth noting a focus on the construction of sewage treatment plants defused competitiveness issues. The group chose to prioritize the reduction of municipal sewage, not industrial sources of pollution. Industries would be connected to the public sewage treatment system, with the result that government (particularly the federal government) and not industrial polluters, would bear the cost of pollution abatement. Thus we see that even though intercantonal cooperation played a role in improving water quality, the cooperation emerged in the context of *federal action*. The efficacy of cantonal cooperation was contingent upon federal legislation to *guarantee* the standards negotiated by cantonal authorities. The availability of federal subsidies for sewage treatment was also crucial. The cantonal officials were no better able than the federal government at tailoring solutions to local conditions, nor in applying the polluter pays principle.

The Role of Referenda in Swiss Environmental Policy

Was Swiss environmental policy effective prior to federal involvement? The ultimate arbiters of Switzerland's environmental policies must be the Swiss electorate. There is substantial evidence of environmental deterioration in

Switzerland, particularly in the period from 1940 to 1965. However, if the Swiss people were untroubled by this situation, then Swiss policy could not be judged inadequate in political or economic terms.

Although opinion polls from the period are lacking, there is evidence that the Swiss were unsatisfied with the state of their environment. Beginning in the nineteenth century, Swiss referendum voters have consistently approved constitutional amendments for greater environmental protection, and often, greater federal powers over environmental protection, at the expense of cantonal power. Between 1971 and 2008, 310 Swiss federal referenda were held. Of these, 23 per cent have had an environmental component (See Table 6.5)

Swiss citizens concerned with environmental issues have often sought to have the matter brought under federal jurisdiction. Most revisions of the Swiss Constitution since 1950 have tended to expand the powers of the federal government (Bussmann 1981, 72). In the case of the environment, constitutional amendments have been proposed when there is a consensus that cantonal policy has been inadequate. The first case was flood control, and by extension, forestry. In the mid nineteenth century, the more mountainous cantons suffered very destructive floods. Expert commissions determined that unsustainable forestry practices were a contributing factor. Although many cantons passed laws on forestry, by 1874 the new constitution granted the federal government powers over forestry and dam building, sufficient to prevent future flooding.

Switzerland is unique in its longstanding and extensive use of referenda, most of which require a double majority for passage. Switzerland has seven types of federal referenda, of which the first four are most relevant for the environment (Table 6.3). Type A are mandatory referenda which are required to approve any amendment to the federal constitution. Type B, the Popular Initiative, is of particular interest. By obtaining enough signatures, a group can force a referendum on a constitutional amendment. This is a uniquely Swiss institution of direct democracy. Type C is the government counter proposal to a Popular Initiative. If the government is opposed to a popular initiative, it can make a formal counter proposal and bring it to the electorate for a vote. The optional referendum, Type D, takes effect if enough signatures are collected on a particular piece of legislation, rather than a constitutional amendment.

Most of the environment referenda have been type A or B. Of the 72 environmental referenda listed below in Table 6.5, 20 were Type A: ordinary constitutional amendments requiring the approval of the electorate. Thirty–one were type B: Popular Initiatives. Five were government counterproposals to Popular Initiatives (Type C). A further 14 were optional referenda on a piece of legislation (Type D). One referendum approved a treaty with Italy and one pertained to a total revision of the constitution.

Table 6.3   Types of referenda used in Switzerland

Code	Type of referendum	Status?	How initiated?	Requirements for passage
A	Constitutional referendum	Obligatory referendum	Automatic upon passage of constitutional amendment by government or, since 1977, membership in international organizations.	Double majority. Majority of population and majority in more than half of cantons (now total of 23 whole cantons.)
B	Popular initiative (also called Constitutional initiative)	Obligatory if conditions are met	Initiative is triggered by proposal from 7 voters and 50,000 signatures (pre–1977) and 100,000 signatures within 18 months (after 1977).	Double majority[1]
C	Counter proposal of parliament to proposed initiative			Double majority
D	Legislative or 'facultative' referendum on a law passed by the federal assembly	Optional	Called if, within 90 days of law's publication, 30,000/ 50,000 voters (pre1977/post1977) or if 8 cantons demand it.	The law is nullified if a simple majority votes against it.[2]
E	Popular initiative for the total revision of the Constitution		100,000 signatures required	Only used in 1872, 1874 and 1935.
F	Treaty referendum	Optional	50,000 signatures	A simple majority is enough for passage.
G	Urgent decree deviating from Constitution			

Notes: [1] Obinger 1998, 251; [2] Linder 1998, 86–7.

Table 6.4 Swiss environmental referenda on expanding federal jurisdiction

Subject	Did it Pass? (Percentage in favour)	Constitutional article created or amended	Date	Type of referendum	Proposed impact on jurisdiction
Flood control and forestry	Yes (63%)	Art. 24	19 April 1874	Popular initiative for total revision (E)	Total revision of the constitution, of which federal authority over flood prevention, and forestry and dam building, was one part.
Streams and forests	Yes (63.5%)	Art. 24	11 July 1897	Constitutional referendum (A)	Gave federal government the authority to supervise forestry as part of flood control and avalanche prevention.
Prevention of water pollution	Yes (81.3%)	Art. 24 quater	6 Dec. 1953	Constitutional referendum (A)	Gave federal government authority to pass laws and regulations about water pollution.
Nuclear energy	Yes (77.3%)	Art. 24 quinquies	24 Nov. 1957	Constitutional referendum (A)	Gave federal government authority over nuclear power.
Highway network	Yes (85%)		6 July 1958	Counter-proposal (C)	Gave Federal Council legal authority to plan a national highway system.
Nature and landscape protection	Yes (79.1%)	Art. 24 sexies	27 May 1962	Constitutional referendum (A)	Gave federal government authority to subsidize projects for protection of nature and the countryside and to legislate on the protection of flora and fauna.
Town and country planning	Yes (55.9%)		14 Sept. 1969	Constitutional referendum (A)	Gave federal government authority to pass framework legislation in this area.
Protection of the environment	Yes (92.7%)	Art. 24 septies	6 June 1971	Constitutional referendum (A)	Gave federal government authority to legislate and regulate on air and noise pollution.
Town and country planning	No (48.9%)		13 June 1976	Optional referendum (D)	Rejected expanded federal role in town and country planning.
Energy policy	No (50.9%)	Art. 24	27 Feb. 1983	Constitutional referendum (A)	Would have granted federal government the authority to make energy policy.
Energy policy	Yes (71%)	Art. 24 octies	23 Sept. 1990	Constitutional referendum (A)	Established constitutional basis for federal energy policy.

The most important referendum in shifting jurisdiction for the environment took place in 1971. As a result of this referendum, the Swiss federal government gained broad powers over many aspects of environmental policy. But referenda in 1897, 1953, 1957, 1962 and 1990 were also important in expanding federal jurisdiction over environmental policy (Table 6.4). Most referenda on extending federal jurisdiction over the environment have passed by a wide margin – the average vote in favour was 70 per cent.

The 1971 environmental referendum was passed by over 90 per cent of the electorate and almost all of the cantons (see Table 6.5 for details). This indicates a very high level of public support for this measure. Similarly, the 1953 referendum that extended federal jurisdiction over water pollution was approved by 81 per cent of voters. The Swiss electorate has rejected an expanded federal role in the environment once, for town and country planning, in 1976. A previous referendum in 1969 authorized the federal framework legislation in this area. Switzerland narrowly voted against federal jurisdiction over energy policy, in 1983, but voted in favour in 1990.

Many of the environment referenda have been popular initiatives, brought as a result of public campaigns collecting a minimum of 50,000 signatures (prior to 1977) or 100,000 signatures for votes held after 1977 (see Table 6.3). Swiss environmentalists have been quite successful in bringing popular initiatives to a vote, but less successful in winning the popular initiatives. Of the 72 environmental referenda listed in Table 6.5, 31 (44 per cent) were popular initiatives, although one was withdrawn before coming to a vote.[8] Of the 31 popular initiatives that came to a vote, only four passed. (One popular initiative sought to weaken environmental protection by raising speed limits. The failure of this initiative represents a victory for environmentalists). The average vote in favour for all popular initiatives was 38 per cent. Many popular initiatives have made fairly radical proposals, such as the 1978 popular initiative to ban car traffic 12 Sundays a year or the 2000 initiative which sought to reduce road traffic by 50 per cent.

On the whole, several conclusions can be drawn from the history of these Swiss referenda. They demonstrate longstanding Swiss concern with environmental issues. The relative prominence of environmental issues among referendum subjects shows that environmental issues are central to the Swiss public: of 310 referenda held between 1971 and 2008, 23 per cent had an environmental component. Although most popular initiatives on the environment have been unsuccessful, the large number of popular initiatives that have been brought on the environment demonstrate the capacity of the Swiss environmental movement.

From 1874, environmental referenda on federal jurisdiction have overwhelmingly extended federal jurisdiction over the environment. If Swiss

---

8   This sonic boom initiative, launched in 1969, never came to a vote because the group withdrew the initiative after the federal government introduced measures which were to the group's satisfaction.

voters were satisfied with cantonal jurisdiction over these many aspects of environmental policy, they would not have voted so frequently and often by very large margins, to grant authority to the federal government.

Conclusion

Switzerland is of interest because it combines a strong historical and institutional predilection for noncentralization, with high levels of concern about the environment. These qualities also characterize the Canadian case to some extent, but Switzerland is a more compelling case because of its institutions of direct democracy. Switzerland's extensive use of referenda has been a counterweight to its traditions of noncentralization and constitutional barriers to federal authority. In Switzerland, evidence from referenda show that the Swiss electorate has not been satisfied with the status quo in environmental protection. In particular, over many decades the Swiss have voted to shift environmental jurisdiction from the cantons to the federal level.

The Swiss demonstrated concern over deteriorating environmental quality, dating from the 1870s. While the cantons had full constitutional authority to act on these problems, individually or cooperatively, the vast majority of cantons proved unable to address these problems. Both pollution that crossed cantonal borders, as well as pollution that was localized within a single canton, were poorly controlled. There is little evidence to support the assertion that cantons carefully tailored environmental policy to local conditions. The inaction by the canton of Valais in the face of over 50 years of highly destructive industrial pollution shows that cantons were unable to control even highly localized pollution problems. Even when there was no competition in the sector from other cantons, the Valais government was extraordinarily reluctant to mandate readily available technical solutions to fluoride pollution.

Where environmental problems did cross cantonal boundaries, the cantons had multiple opportunities for cooperation. They could have concluded formal cooperative agreements, using the Konkordat tool. Or they could have employed informal cooperation, through intercantonal working groups, which are ubiquitous in most areas of Swiss policy. However, prior to a stronger federal role, almost no environmental cooperation took place. Cantonal working groups were only able to agree upon water pollution control measures once they were certain these would be embodied in federal legislation. In order to avoid regulating industrial pollution, they reached agreement over sewage treatment plants. Even here, the availability of federal subsidies for the construction of sewage treatment plants was decisive in the exponential growth in sewage treatment plant construction that followed federal encroachment into this area of jurisdiction.

This unwillingness or inability to act persisted until the federal government began to extend authority over pollution control in the late 1950s, subject to approval through referenda. Through several referenda, beginning in 1897, the

Swiss electorate has approved expanded federal authority over the environment, starting with forestry and flood control and ending most recently with energy policy. If the Swiss electorate opposed these shifts in jurisdiction, they had ample opportunity to retain the status quo of cantonal control by voting against these referenda. But they did not.

It is puzzling that the cantonal governments were less able to satisfy people's preferences for environmental quality than the larger federal government. In theory, smaller democratic units are more responsive to popular preferences than larger ones. However, this assumption does not hold if there is a collective action problem. The cantons' inability to deliver on the level of environmental protection desired by the populace persisted for decades. Environmental regulation represented a collective action problem which could not be overcome through cooperation, whether formal or informal. If the very 'green' Swiss have been unable to provide adequate environmental protection on a noncentralized basis, it seems unlikely that anyone else should be more successful in this endeavour.

Table 6.5 Swiss referenda with an environmental component

Subject	Date	Type of referendum	Success or failure	Yes Vote		Per cent turnout
				Per cent of voters	Number of Cantons	
Flood control and forestry	19 April 1874	E	S	63.0	13.5	N/A
Streams and forests	11 July 1897	A	S	63.5	16.0	38.5
Legislation on water powers	25 Oct. 1908	C	S	84.4	21.5	48.2
Navigation	4 May 1919	A	S	83.6	22.0	53.9
Automobiles and cycles (art. 37)	22 May 1921	A	S	59.8	15.5	38.5
Subsidy of Alpine Roads (art. 30)	15 May 1927	A	S	62.6	21.0	55.3
Traffic law for cars and bicycles	15 May 1927	D	F	40.1	–	57.8
Federal traffic law	12 May 1929	B	F	37.2	3.0	65.0
Diversion of traffic	5 May 1935	D	F	32.3	–	63.2
Transport by automobile	25 Feb. 1951	D	F	44.3	–	52.4
Agriculture Act	30 Mar. 1952	D	S	54.0	–	64.1
Prevention of water pollution (art. 24 quater)	6 Dec. 1953	A	S	81.3	22.0	59.1
Protection of Rhine up to Rheinau	5 Dec. 1954	B	F	31.2	1.0	51.9
Utilization of hydroelectricity	13 May 1956	B	F	36.9	2.5	52.1
Nuclear energy (art. 24 quinquies)	24 Nov. 1957	A	S	77.3	22.0	45.5
Authorizing Federal Council to plan a national highway system	6 July 1958	C	S	85.0	21.0	42.4
Ratification of treaty with Italy over use of the Spöl River	7 Dec. 1958	F	S	75.2	–	46.4
Taxes on motor fuel	5 Mar. 1961	D	F	46.5	–	63.3
Nature and landscape protection (art. 24 sexies)	27 May 1962	A	S	79.1	22.0	38.7
Town and country planning	14 Sept. 1969	A	S	55.9	19.5	32.9

Table 6.5 continued

Subject	Date	Type of referendum	Success or failure	Yes Vote		Per cent turnout
				Per cent of voters	Number of Cantons	
Regulation of sonic booms	1971	B Popular initiative started in 1969		Satisfactory regulations were agreed upon in negotiation with the government and the initiative was withdrawn after 15 months		
Protection of the environment (art. 24^{septies})	6 June 1971	A	S	92.7	22.0	37.9
Protection of animals	2 Dec. 1973	A	S	84.0	22.0	35.0
Legislation on water resources	7 Dec. 1975	A	S	77.5	21.0	30.9
Town and country planning	13 June 1976	D	F	48.9	–	34.6
Air pollution from cars	25 Sept. 1977	B	F	39.0	1.5	51.7
Democracy in highway construction	26 Feb. 1978	B	F	38.7	0.0	48.2
Twelve Sundays a year without motor traffic	28 May 1978	B	F	36.3	0.0	48.8
Protection of animals	3 Dec. 1978	D	S	81.7	22.0	43.3
Pedestrian trails	18 Feb. 1979	C	S	77.6	22.0	49.6
Nuclear plants	18 Feb. 1979	B	F	48.8	9.0	49.6
Revision of atomic energy	20 May 1979	D	S	68.9	22.0	37.6
Constitutional article on energy	27 Feb. 1983	A	F	50.9	11.0	32.4
Gasoline tax	27 Feb. 1983	A	S	52.7	15.5	32.4
Tax on trucks	26 Feb. 1984	A	S	58.7	15.5	52.8
Highway tax	26 Feb. 1984	A	S	53.0	16.0	52.8
End atomic development	23 Sept. 1984	B	F	45.0	6.0	41.7
Taxation of heavy trucks	7 Dec. 1986	B	F	33.9	0.0	34.0
'Rail 2000' project (to improve Swiss railway system)	6 Dec. 1987	D	S	56.7	18.5	47.7

Table 6.5 continued

Subject	Date	Type of referendum	Success or failure	Yes Vote Per cent of voters	Yes Vote Number of Cantons	Per cent turnout
Stop Rothenturm military base to protect moors and wetlands	6 Dec. 1987	B	S	57.8	20.0	47.7
Coordination of transport'n policy (two constitutional articles)	12 June 1988	A	F	45.5	3.5	41.9
Protection of small farms	4 June 1989	B	F	48.9	8.0	36.0
Raise highway speed limit to 130km/h	26 Nov. 1989	B	F	38.0	5.0	68.6
Stop all new road construction	1 April 1990	B	F	28.5	0.0	40.5
Prevent highway between Murten and Yverdon	1 April 1990	B	F	32.7	0.0	40.5
Prevent highway between Wettswil and Kronau	1 April 1990	B	F	31.4	0.0	40.5
Prevent highway between Biel and Solothurn	1 April 1990	B	F	34.0	0.0	40.5
Establish constitutional basis for federal energy policy	23 Sept. 1990	A	S	71.0	23.0	39.6
10 year moratorium on nuclear plant construction	23 Sept. 1990	B	S	54.6	19.5	39.6
End use of nuclear energy	23 Sept. 1990	B	F	47.1	7.0	39.6
Transfer funds from road construction to public transportation	3 March 1991	B	F	37.1	1.5	31.1
Reduce amount of water in reservoirs to protect the environment	17 May 1992	B	F	37.1	0.0	38.4
Less drastic reduction than in previous proposal	17 May 1992	D	S	66.1	18.5	38.5
Regulation of genetic technology	17 May 1992	D	S	73.8	22.0	38.0

Table 6.5 continued

Subject	Date	Type of referendum	Success or failure	Yes Vote Per cent of voters	Yes Vote Number of Cantons	Per cent turnout
Raise gasoline tax by 20 cents per litre	7 Mar. 1993	D	S	54.5	15.0	51.3
Protect the Alpine region from transit traffic	20 Feb. 1994	B	S	51.9	16.0	40.9
Continuation of the heavy goods vehicle tax	20 Feb. 1994	A	S	72.0	23.0	40.8
Continuation of the motorway tax	20 Feb. 1994	A	S	68.5	21.0	40.8
Engine size related heavy goods vehicle tax	20 Feb. 1994	A	S	67.0	21.0	40.8
To protect life and the environment from gene manipulation	7 June 1998	B	F	33.0	0.0	41.3
Engine size related heavy goods vehicle tax	27 Sept. 1998	D	S	51.8	–	51.8
Town and country planning	7 Feb. 1999	D	S	55.9	–	38.0
Traffic halving initiative	12 Mar. 2000	B	F	21.3	0.0	42.4
Initiative for renewable energy levy	24 Sept. 2000	B	F	31.2	0.0	44.7
Counter–proposal to renewable energy initiative	24 Sept. 2000	C	F	45.3	4.5	44.7
Counter–proposal to withdrawn 'Energy Environment' initiative	24 Sept. 2000	C	F	44.5	2.5	44.9
'Roads for everyone' initiative (reduced speed limits)	March 4, 2001	B	F	20.3	0	55.8
Four car–free Sundays per year	18 May 2003	B	F	37.7	0	49.8
Gradual decommissioning of nuclear power plants	18 May 2003	B	F	33.7	0.5	49.7
For an extension of moratorium on nuclear power plant construction	18 May 2003	B	F	41.6	1	49.6
'For food grown without genetic modification'	27 Nov. 2005	B	S	55.7	23	42.2
'Against the noise of fighter jets in tourist areas'	24 Feb. 2008	B	F	31.9	0	38.7

Source: Center for Research on Direct Democracy 2008.

Chapter 7

Canada: When Centralization Does Not Occur

... Canada's environmental record is among the best in the world ...

Environment Canada, *State of the Environment Report – 1996*

Canada's environmental performance is, by most measures, the worst in the developed world. We've got big problems.

Prime Minister Stephen Harper, television interview December 2006 (Hahn 2006)

It is unusual for heads of state to publicly declare their countries to be 'the worst' in anything. Furthermore, Prime Minister Harper's tone was in stark contrast to that adopted by his predecessor Paul Martin, 12 months earlier, at a Montreal conference on the Kyoto protocol. At that conference, then Prime Minister Martin adopted a superior tone, scolding the US on climate change: '[t]o the reticent nations, including the United States, I say there is such a thing as a global conscience and now is the time to listen to it' (Revkin 2005). How can the above statements be reconciled?

Prime Minister Harper was closer to the mark than Prime Minister Martin. Prime Minister Martin was engaged in pre–election pandering, appealing to Canadians' self–perception of moral superiority, rather than speaking from a position of environmental leadership. When he lectured the US, Prime Minister Martin knew very well that since the Kyoto protocol was signed, US greenhouse gas emissions had risen 12 per cent. However, Canada's greenhouse gas emissions had increased 24 per cent in the same time period. Thus Canada was moving ever farther away from compliance with Kyoto, despite the fact that the US had not ratified Kyoto and Canada had.

While Martin's 2005 statement was deliberately misleading, the 1996 claim can be attributed to ignorance. A major consequence of the decentralized nature of Canadian environmental policy is a critical lack of information. It has been very difficult to obtain and compare information on environmental quality and environmental regulations in Canada. Interprovincial and international comparisons have been difficult, if not impossible, to make. It seems likely that Canadian environmental policy improved (or at least stood still) between 1996 and 2006. But the 1996 statement no longer holds true because the evidence that has since become available contradicts that noble claim.

Canadian environmental quality and environmental policy are worse than one might expect in a relatively wealthy country. The argument here is that the decentralized character of Canadian environmental policymaking has resulted in lower levels of environmental protection. This is so, despite evidence that Canadians care at least as much about the environment as their neighbours to the south. The Canadian case represents a counterfactual to those who would argue that environmental policy in the United States and Switzerland would have become stringent over time anyway, without the increased federal role observed in both of those countries. The Canadian case offers compelling evidence that the improvements observed in the US and Switzerland cannot be explained primarily by changes in public opinion, because Canadian public opinion also strongly supports environmental protection.

This book argues that the lack of centralization of Canadian environmental policymaking is the main cause of Canada's poor environmental performance. The ten provinces have primary responsibility for the environment, particularly the regulation of industry, because of the division of powers in the Constitution of 1867. Subsequent Supreme Court decisions suggest that an expanded federal role would survive a legal challenge. However, in recent times, the obstacles to federal action are as much political as constitutional. Canada's recurrent state of constitutional crisis makes the federal government wary of offending the provinces, which jealously protect all their areas of jurisdiction.

If Canadians are so concerned about the environment and the provinces do not want the federal government to interfere, why do the provinces not simply cooperate? The federal and provincial governments have been meeting and discussing the environment for decades, both formally and informally.[1] There is little to suggest that the environment has become significantly cleaner as a result. The unanimous agreements produced are not binding and there is little monitoring. Provinces often do not adopt as their provincial standards the guidelines unanimously agreed to in negotiations. The agreements are vague and not subject to enforcement, in the way that many international treaties are. The failure to negotiate environmental agreements with targets for performance suggests that cooperative agreements are not a viable vehicle for addressing environmental problems.

This outcome is consistent with Fritz Scharpf's description of a 'joint decision trap' in German federalism and the European Union. When actors collectively make decisions, subject to unanimity rules (also known as unit veto), the resulting decisions tend to be at the lowest common denominator. Scharpf concluded that the joint decision trap is

> ... an institutional arrangement whose policy outcomes have an inherent (nonaccidental) tendency to be suboptimal – certainly when compared to the policy potential of unitary governments of similar size and resources. Nevertheless, the arrangement represents a 'local optimum' in the cost benefit calculations of all participants that might have the power to change it. If that is

---

1   The first documented conference at the Ministerial level was the National Conference on Pollution and our Environment, held by the Canadian Council of Resource Ministers in 1966.

so, there is no 'gradualist' way in which joint–decision systems might transform themselves into an institutional arrangement with greater policy potential (Scharpf 1988, 271).

The description of a suboptimal, yet highly stable system, is a fair characterization of Canadian environmental policy, and some other areas of Canadian policy such as internal trade.

This chapter will demonstrate that, despite high levels of public concern, Canada's record on environmental protection is quite poor. It goes on to argue that this situation is attributable to the noncentralization of environmental policy in Canada. Furthermore, cooperation has done little to improve outcomes. The chapter is divided into nine sections. The first section presents data on several measures of environmental quality. The second section surveys Canadian public opinion on the environment. The third describes the constitutional division of powers in Canada and the federal government's reluctance to use some of its powers. The fourth sets out the Canadian federal government's limited role in environmental policy. The fifth section describes the federal government's repeated attempts to spur provincial action. The sixth sets out the patchwork of provincial and federal legislation which comprises the Canadian regulatory system. The seventh section examines theoretical claims about decentralization, in light of the Canadian case. The eighth section characterizes the Canadian experience with interprovincial cooperation, including the Harmonization Accord and setting Canada–wide Standards. The ninth section concludes.

The Quality of Canada's Environment

Environmental degradation is very much at odds with Canada's image as a vast expanse of pristine wilderness. Although Canada has very low population density (three persons per km²), over 75 per cent of the population live in urban areas, a higher proportion than in the US. More than half of Canadians live in a narrow corridor along the US border, between Windsor ON and Quebec City PQ. This corridor has also been the location of much heavy industry, although pulp and paper mills and mines often lie in the hinterlands. Three quarters of Quebec's industries lie between the Ontario border and the town of Sorel, 40 miles downriver from Montreal (Canada 1991, 19–9). Thus, while Canada is an enormous country, its pollution is not evenly dispersed. Rather, it is concentrated in the most heavily populated regions. Because the regions of greatest population and pollution coincide, pollution in Canada affects people and their quality of life, not just ecosystems.

Until quite recently, it has been difficult to create a national picture of environmental quality in Canada. This is another result of decentralization. The OECD's 2004 *Environmental Performance Review* for Canada characterizes environmental information as incomplete and 'surprisingly poor in some areas'(OECD 2004A, 148). The federal government halted publication of its *State of the Environment* reports in 1996, due to budget cuts (OECD 2004A, 148).

Many provinces have not collected environmental quality data. What provincial data exists has been compiled in incompatible ways, making it very difficult to make interprovincial comparisons or to aggregate data in order to assess the situation for Canada as a whole (OECD 2004, 58). For example, the *first* national assessment of water quality was released in 2003, based on data from 319 sampling stations. In 1993, with the creation of the National Pollutant Release Inventory, it became possible for the first time to systematically track emissions of toxic substances from Canadian industry.

In 2005, the federal government resumed the analysis and publication of the Canadian Environmental Sustainability Indicators, including greenhouse gas emissions, air quality and water quality. The 2007 report presented data up to 2005. By combining these findings with the results of the OECD's second *Environmental Performance Review of Canada* (2004a), it is possible to discern some trends. Although there have been some improvements, many of the trends show no improvement or deteriorating environmental quality.

Trends: Greenhouse Gas Emissions

One area in which Canada has performed particularly poorly and continues to do so is in greenhouse gas reduction. Although Canada contributes only two per cent to global emissions, Canadians' per capita emissions are among the highest in the developed world. In addition, the greenhouse gas intensity of Canada's economy is greater than that of other OECD countries. Canada's high per capita emissions suggest that there may be many opportunities for reductions, overlooked improvements in efficiency which could easily reduce emissions. Under the Kyoto protocol, Canada committed to reducing GHG emissions by six per cent (by 2008–2012), over the 1990 baseline. However, between 1990 and 2005, Canada's GHG emissions increased 25 per cent. As a result, by 2005, Canadian annual emissions were 33 per cent above Canada's Kyoto target (Environment Canada et al. 2007, 3).

Much of the increase in emissions came from the extraction, processing, refinement and transportation of oil and gas (Environment Canada 2007a). Developing of Alberta's oil sands, as a source of petroleum, has contributed to GHG emissions, particularly since exploitation of the oil sands emits considerably more greenhouse gases per barrel of oil than the exploitation of conventional oil resources. The disparity of interests between oil producing provinces and those with a less greenhouse gas intensive economies makes national agreement on greenhouse gas reductions very difficult (Weibust 2003).

Trends: Air Quality

Canada has seen some improvements in air quality. Since the OECD's first assessment in 1995, sulphur oxides (SO_x) emissions declined, due in part to changes resulting from cooperative Canada–US efforts to reduce acid rain. Ambient levels of nitrogen oxides (NO_x), SO_x, carbon monoxide and Total Suspended Particles decreased in urban areas (OECD 2004a, 43). Between 1990 and 2001, smog precursors NO_x and Volatile Organic Compounds (VOCs) showed modest declines, 6 per cent and 17 per cent respectively.

The decline in NO_x fell short of the target negotiated by the Canadian Council of Ministers of the Environment (CCME) in the 1990 *Canada NO$_x$/VOCs Management Plan.* It set a target of an 11 per cent reduction by 2005, over 1985 levels. Some of the reduction in NO_x/VOCs is attributable to more stringent standards for vehicles. Due to the integrated North American automotive market, Canadian vehicles largely comply with US Environmental Protection Agency standards for emissions, which accounts for part of the decline (Olewiler 2006, 134–5). Other air pollutants show deterioration. Between 1990 and 2005, Canadians' exposure to ground level ozone increased by 12 per cent. The increases occurred primarily in southern Ontario and Quebec; other regions showed no change (Environment Canada et al. 2007, 2).

However, despite some progress on air quality, the 2004 OECD *Review* observed that levels of 'emissions of traditional air pollutants in Canada remain very high compared with most OECD countries' (OECD 2004a, 33). Measured per capita and also per unit of GDP, emissions of traditional air pollutants (e.g. SO_x, NO_x and VOCs) in Canada were among the highest in any OECD country. Canada's largest source of SO_x emissions was industry, with almost half of industrial emissions coming from non–ferrous mining and smelting (OECD 2004a, 36). Canadian sulphur dioxide emissions per unit of GDP were twice the OECD average and 30 per cent higher than US levels (OECD 2004a, 37). The report indicates that Canadian economic activity is relatively pollution intensive, in comparison to the US and the other OECD countries.

Trends: Water Quality

It has been very difficult to assess trends in Canadian water quality, due to lack of data. The Canadian Environmental Sustainability Indicators provide a snapshot of fresh water quality from 2003–2005, indicating if the water can sustain fish. The Indicators showed that, of 359 freshwater sites in southern Canada, only 6 per cent had excellent water quality. Thirty–eight per cent of sites had good water quality, with a further 33 per cent showing fair water quality. The remaining 23 per cent of sites in southern Canada had marginal or poor water quality (Environment Canada et al. 2007, 4).

Measuring Industrial Emissions

In 1993, the creation of a Canadian inventory of toxic releases provided an unprecedented opportunity to assess emissions to the environment and pollution by industry in Canada. Prior to the release of this data, it was very difficult to assess the pollution performance of Canadian firms. The US Toxics Release Inventory (TRI), upon which the Canadian National Pollutant Release Inventory (NPRI) is based, began collecting toxic release data in 1987.[2] The Canadian inventory, lists 230 substances and requires facilities with 10 or more employees using NPRI–listed substances in concentrations >1 per cent and in quantities greater than or equal to 10 tonnes to report transfers and releases to Environment Canada (Olewiler and Dawson 1998). The NPRI makes a distinction between releases and transfers. Releases are on–site discharges to air, water or land, whereas a transfer is the shipment of waste to an off–site location, which can include landfill, incineration and municipal sewage treatment.

In 1997, the Commission for Environmental Cooperation (CEC) released *Taking Stock: North American Pollutant Releases and Transfers*. This was the first comprehensive comparison of industrial emissions in Canada and the US. It showed that Canadian factories polluted far more than their US counterparts. The study examined emissions in 1994 from 1,707 Canadian factories and 22,744 American factories, using data drawn from national toxic release inventories.[3] The study found that, for some types of industrial pollution, Canada produced more pollution than the US *in absolute terms* i.e. all of Canada produced more than all of the US, (even though the US population and economy are ten times the size of Canada's). The disparity was especially strong in water pollution. Canadian facilities accounted for 7.4 per cent of the total number of facilities reporting in both Canada and the US total, but these Canadian facilities accounted for 35.8 per cent of the total discharges to surface water.

The report found that the average releases and transfers from each facility in the US were 58,358 kg and in Canada were 118,414 kg – more than twice as much. The report notes:

> This significant difference does not appear to arise from the average number of [chemicals] reported by each facility, from differences predominating in the use of chemicals at NPRI versus TRI facilities or from differences in reporting thresholds between the two [inventories] (CEC 1998, 55).

The report matched industrial categories (using US Standard Industrial Classification codes) between the two inventories, in order to provide a more valid comparison.

In 15 of the 20 matched categories, average releases and transfers were higher for Canadian facilities. This included the categories with the largest total releases in both countries: paper products, chemicals and primary metals industries. These

---

2   In the US, a facility with 10 or more full–time employees must report for the TRI if it manufactures or processes more than 25,000 pounds or uses more than 10,000 pounds of any listed chemicals during the calendar year (equivalent to 11.3 and 4.5 tonnes respectively).

3   'Canada Outspews the US: Average Plant Here Pollutes More Than Twice as Much' 1997, A1.

industries accounted for 90 per cent of releases to surface water and transfers to sewage treatment plants in each country. Thus the sectors which tend to be the biggest polluters produced more pollution on average in Canada than they do in the US.

In Canada, the paper industry accounted for 77 per cent of all releases to surface water in 1995. When four large facilities each discharging more than 1.5 million kg are excluded, the remainder of the Canadian paper industry made up 45 per cent of releases to surface water. What is remarkable is that releases from the Canadian paper industry had declined in the 1990s, thus these releases must have been substantially higher prior to the 1995 report. The report noted that 'the Canadian pulp and paper industry reported reductions in surface water discharges of 15 per cent from 1994 to 1995, *despite a 14 per cent increase in the number of reporting facilities* ... the [industry] is projecting a 38 per cent decrease in releases and transfers from 1995 to 1997'. This strongly suggests that average releases were substantially greater in the 1970s and 1980s.

Canadian economist Nancy Olewiler used the pollutant release inventories to compare toxic releases per job, and also per dollar of output for Canada and the US. Her analysis showed that the toxic releases both per job and per dollar of output were 50 per cent greater in Canada for the following industries: chemicals and chemical products, nonmetallic minerals, paper and allied products, refined petroleum and coal, and rubber and plastics. For Canadian releases, she also computed a *toxic intensity ratio* that weighted the quantity of emissions by measures of the toxicity of the compounds being released. She concluded that:

> The Canadian industries that perform least well in controlling emissions are those that have the greatest potential for causing environmental damage, as shown with our toxic intensity indicators. Overall, emissions per job and per dollar of output from Canadian manufacturing industries are 50 percent higher than releases from US manufacturing industries (Olewiler and Dawson 1998 ,7).

These figures suggest that Canadian environmental regulations have been far from optimal: the industries whose emissions are *most* toxic are the most pollution intensive; that is, they pollute more per unit of output and employment. The analysis also indicated that, despite Canadian politicians' concern about employment, Americans got more jobs per unit of pollution than Canadians did.

Although these findings are quite damning, they may understate Canadian toxic releases *vis–à–vis* the US. During this time period, the Canadian registry covered only 176 substances whereas the US TRI covered 606 substances. Also, the threshold at which Canadian facilities were required to report was higher: '[a]lmost 16 times as many industrial facilities reported to national pollution reporting bodies in the United States as did their counterparts in Canada' (Fine 1997, A4). However, the American TRI covered only manufacturing industries whereas the Canadian NPRI also includes primary industries.

In 1992, it was possible for a Canadian environmental politics expert to say:

... Canadians have nothing to be ashamed of; this comparison suggests that the
Canadian environmental record compares favourably with the American record
in some areas (Hoberg 1992, 261).

In light of the data presented by the Commission for Environmental Co–operation,
it is difficult to make this claim. Where industrial pollution is concerned, Canadian
factories emitted far more pollution than comparable American ones.

The available data suggests that many types of industrial pollution have
declined since the Commission on Environmental Cooperation began its analysis
of toxic releases in the 1990s. However, although Canadian emissions have
declined, the US emissions have declined more rapidly. A 2005 comparison of
NPRI and TRI data show that, between 1995 and 2003, US toxic emissions to
the air declined by 45 per cent, whereas Canadian emissions during this period
declined by only 1.8 per cent (Sallot 2005, A7). Thus while Canadian emissions
show some improvement, the gap between Canadian and American emissions is
not narrowing significantly because the Americans are reducing emissions faster.

Municipal Sewage Treatment

Canada's poor environmental performance is not limited to industrial sources.
Canada's environment is also blighted by human waste in untreated or inadequately
treated sewage. Canadian water quality scientists have concluded: 'there is a need
to review sewage treatment requirements in Canada'(Chambers et al. 1997, 659).

The first *National Sewage Report Card* issued in 1994 by the Sierra Legal
Defence Fund describes sewage treatment in Canada as a 'national disgrace'(Bonner
1994). In 1994, three provincial capitals, Halifax NS, St. John's NL and Victoria
BC, had no sewage treatment whatsoever. Together they discharged 105 billion
litres of raw sewage into the oceans each year. In theory, provincial legislation and
the federal *Fisheries Act* prohibit such releases of sewage. Of the twenty cities in
the study, eleven violated provincial permits or held no permits whatsoever.

Ten years later, the Sierra Legal Defence Fund examined 22 major cities in its
third *National Sewage Report Card*. Since its second report card in 1999, fourteen
cities had made some progress, five had made no progress and three cities had
regressed. As of 2004, Victoria BC was the only city in Canada that continued to
discharge *all* of its sewage without any treatment, dumping more than 120 million
litres of raw sewage daily into the ocean. Saint John NB, Saint John's NL and
Halifax NS continued to discharge some of their sewage raw and untreated (Sierra
Legal Defence Fund 2004).

Even in this undistinguished company, Montreal stood out for decades for its
exceptionally bad performance. As early as 1909, the journal *Sanitary Review*
described the city's wastewater disposal as 'a hygienic disgrace to civilization'
(Benidickson 2007, 179). By the 1980s, Montreal, a city of almost two million, was
still without any sewage treatment. Unlike Victoria BC, which dumps its sewage
into the ocean, Montreal discharged its sewage to the St. Lawrence River, from

which 45 per cent of Quebec's population drew its drinking water (Canada 1991, 19–1). Furthermore, Montreal is quite far upstream on the St. Lawrence, exposing a large proportion of communities on the river to bacterial contamination.

In the 1990s, Montreal began to fulfill commitments to sewage treatment it had made in 1975 (Bonner 1994, 23). By 2000, 100 per cent of Montreal's sewers were connected to sewage treatment plants. However, these plants only provided *primary* treatment, which removes about 40 per cent of Biological Oxygen Demand (BOD) and about 60 per cent of Total Suspended Solids (TSS) (Wristen 1999, 15). (The minimum standard required by the US and European Union is secondary treatment). Primary treatment does not eliminate bacteria in the water discharged by the sewage treatment plant. In 2008, the city approved a $200 million project to disinfect effluent from its sewage treatment plant.[4]

It could be argued that these levels of sewage treatment reflected public preferences in Quebec. Opinion polls suggest otherwise: a 1981 poll of 2000 Canadians showed that 43 per cent of Quebecers identified water pollution as the most important environmental problem at the provincial level (see Table 7.1).[5] They ranked the problem above acid rain, the second highest response (10 per cent), which has more significant environmental impacts in Quebec than in other provinces. This response does not reflect a general preoccupation with water pollution: When asked about environmental problems at the *national* level, Quebecers were not significantly more concerned about water pollution than any other Canadians. Similar results were obtained when the same question was asked in 1982.[6]

Table 7.1 Public perception of most important environmental problems at the provincial and national levels, in per cent

1981	Water Pollution		Acid Rain		Air Pollution	
	Provincial	National	Provincial	National	Provincial	National
Western Canada	26	22	6	18	8	5
Ontario	21	23	27	18	11	12
Quebec	43	26	10	14	6	10
Atlantic Canada	11	13	12	24	6	9
Canada	27	22	15	17	8	9

Source: CROP Survey Report 81–4 (Bird and Rapport 1986, 256).

---

4   'Committee Okays Cleansing Project' 2008, A7.
5   CROP Survey Report, 81–4 (Bird and Rapport 1986, 256).
6   CROP Survey Report, 81–4 (Bird and Rapport 1986, 256).

Nor are Canadian cities with sewage treatment facilities above reproach, particularly in international comparison. Of the 20 major Canadian cities studied in the 1994 *National Sewage Report Card*, 17 discharged some untreated sewage every day. Only two of the 20 cities used tertiary treatment to treat a substantial proportion their sewage. In contrast, by the 1990s, over 90 per cent of the population of Sweden and Finland were connected to tertiary sewage treatment facilities. The only Canadian jurisdiction that even approaches this percentage is Ontario, where 60 per cent of the population was connected to tertiary treatment. In 1991, four provinces had no tertiary treatment facilities whatsoever (Canada 1991, 3–9). By 1994, 34 per cent of Canadians had tertiary sewage treatment (Chambers et al. 1997, 662).

The majority of Canadian cities have sewers from the 1940s or earlier that combine sanitary sewers for sewage, with storm sewers, for rainfall. In combined sewers, rain water mixes with raw sewage and is discharged directly into a lake or river, termed combined sewer overflows (CSO). The governments of United States and Germany have made extensive efforts to reduce pollution through CSOs. After a US federal court ordered a water cleanup, the metropolitan Boston area invested billions of dollars in upgrading sewers and sewage treatment, in part to reduce or eliminate water pollution from combined sewer overflows (Rogers 1996, 96). Canada has not made similar investments and the OECD 2004 *Review* identified this as a priority for improvement.

Discharging untreated sewage may appear to save money, but this is a shortsighted view. Epidemiological studies in Montreal in 1991 and 1997 found that contaminated tap water accounted for between 14 and 40 per cent of gastrointestinal illnesses (Boyd 2007, 16). Sewage pollution has imposed measurable costs in lost recreational opportunities and contaminated fish and shellfish. Sewage contamination of lakes and rivers deprive millions of Canadian city dwellers of the use of easily accessible urban beaches.

The economic cost of sewage pollution has been felt directly by the shellfish and aquaculture industries. On the Atlantic coast (excluding Quebec), 35 per cent of the areas deemed suitable for direct harvesting of shellfish were closed in 1995, due to high bacterial counts, a loss of $10 to $12 million to the local economy. In Quebec, of 156 shellfish zones evaluated in 1991, 70 per cent were closed conditionally or permanently due to bacterial contamination. In Newfoundland, 30 per cent of proposed aquaculture sites were rejected in 1989 because of high coliform bacteria levels, indicating the presence of untreated sewage. In British Columbia, sewage effluent was estimated to be the sole cause of 15 per cent of shellfish harvesting closures and a contributing factor in an additional 78 per cent of closures (Chambers et al. 1997, 687–8). These problems persisted over time: about one quarter of the almost 20 000 km^2 monitored in 2002–03 under the Shellfish Water Quality Protection Programme were classified as closed, primarily due to bacterial contamination (OECD 2004a, 60).

While Canadian environmental quality has shown improvement in some areas, it has been and remains poor in comparison with other OECD countries, particularly the US. One possible explanation for Canada's poor environmental performance might be that Canadians do not care about the environment. Although opinion poll data are limited, especially prior to the 1980s, the available data does not support the conclusion that Canadians are unconcerned about the environment or that they care less than Americans do.

Canadian Public Opinion on the Environment

In Canada, pollution emerged as a political issue at about the same time it did in the US, peaking in the early 1970s. In a 1970 Canadian Gallup poll, 70 per cent of respondents stated that pollution was a serious national problem. Another poll that year found that respondents rated the environment 'the most important problem government should be addressing'. By 1971, the percentage expressing concern had dropped by 50 per cent (Parlour and Schatzow 1978, 13). There was a second peak of concern in the late 1980s. Although levels of concern have risen and fallen over the last three decades, public concern about the environment has generally remained high, on par with the level of concern expressed in the US. Furthermore, Canadians are more likely than Americans to favour government action to address a variety of issues. It is all the more puzzling then, that the US government has been more aggressive and effective than the Canadian in addressing pollution.

As far back as 1970, there is evidence that Canadians were unwilling to give industry carte blanche, despite concerns about potential job losses. In 1970, 700 residents of Hamilton, a polluted Ontario steel town, were asked about pollution and unemployment (Winham 1972, 389–401). In general, levels of environmental concern in Hamilton were higher than for the rest of Canada. When a sample of Canadians was asked to identify the three most important problems, 32 per cent of those polled identified pollution and 25 per cent said unemployment (the two most popular responses). When the same question was asked in Hamilton, 41 per cent of the population chose pollution and 17 per cent picked unemployment. Only 10 per cent of the sample said they were unwilling to pay more taxes in order to reduce pollution of Lake Ontario caused by Hamilton's sewage system. Cross tabulating these results with respondents who favoured continued industrial development in Hamilton reveals that even those who favoured more growth were 79 per cent in favour of such a tax.

Polling data from the early 1980s showed that Canadians were strongly in favour of more government control over environmental standards. In 1980 and 1982, 86 per cent and 80 per cent respectively stated that they favoured 'greater government control over environmental standards' (Johnston 1986, 213). This was one part of a larger question on several issues; only 'Health and Safety Standards' elicited a higher response in favour of government control. When 1980 responses were cross–tabulated by responses to a question on government regulation in general, even those favouring 'less regulation' were 88 per cent in favour of more government regulation of environmental standards (Johnston 1986, 214).

More recent data (1990 and 1993) show Canadians' concern for the environment was either equal to or greater than that of Americans surveyed. The 1990 World Values Survey contained six questions on environmental matters. On five of these questions, the Canadians' and Americans' responses were either identical or within two per cent (as a percentage of those agreeing with the statement in the question) (Inglehart et al. 1998, V12–V16). On several questions about willingness to pay more for environmental protection, Canadians and Americans showed greater willingness than West Germans. However, West Germans were less likely (21 per cent) than Canadians and Americans (30 per cent; 31 per cent) to agree that environmental problems had to be accepted in order to combat unemployment (Inglehart et al. 1998, V16).

The exception on Canadian/American agreement was a question asking respondents if the urgency of protecting the environment was exaggerated. Twenty–nine per cent of US respondents agreed or strongly agreed with this statement, while only 22 per cent of Canadians polled did. Only 12 per cent of West Germans agreed or agreed strongly with the statement (Inglehart et al. 1998, V17).

In 1993, the International Social Survey Programme (ISSP) asked the citizens of 20 countries about their opinions on the environment (Frizzell and Pammett eds. 1997). The survey examined the accuracy of people's environmental knowledge, their concern about the environment, their perception of hazards, their activism and their preferences for paying for the environment. With regard to the accuracy of knowledge, Canadians ranked first of all the countries surveyed. East Germany, the US and West Germany were ranked 6th, 8th and 10th respectively. The only country in the sample which was consistently more 'green' than Canada in its opinions was the Netherlands.

The survey also sought to identify the respondent's ideological position. The results were consistent with the previous literature suggesting that Canadians are more in favour of government intervention and generally less individualist than Americans are (Lipset 1990). The data in Table 7.2 show that Canadians believed more strongly in government's role in income redistribution than Americans did.

This Canadian belief in a strong role for government carried over to environmental policy. Canadians polled did not believe that voluntary action by businesses and individuals was sufficient to protect the environment (Tables 7.3a and 7.3b). They favoured a strong role for government, even if this restricted the freedom of businesses and individuals. The Canadian responses were closer to those of the Germans than the Americans with regard to their willingness to limit the rights of business.

The ISSP survey asked Canadians and Americans about several potential environmental hazards (Clarke and Stewart 1997) (see Table 7.4). They were asked about six types of pollution, assessing the dangerousness of each with regards to (1) themselves and their families and (2) the general environment. For each of the

12, Canadians deemed each type of pollution to be somewhat more dangerous than the American respondents did. When the survey asked about willingness to pay, either higher prices or higher taxes for environmental protection, majorities in both countries were prepared to do so and sizeable minorities in both countries were prepared to accept lower living standards (US 34 per cent; Canada 44 per cent) in Table 7.5.

Table 7.2 Polls on the role of government; the environment versus the economy

Questions asked	Percentage who strongly agree or agree			
(The 6 possible replies are strongly agree, agree, neither agree nor disagree, disagree, strongly disagree and can't choose)	Canada	United States	West Germany	East Germany
Private enterprise is the best way to solve economic problems	51	49	64	52
It is the responsibility of the government to reduce the differences in income between people of high incomes and those with low incomes	42	30	56	61
We worry too much about the future of the environment and not enough about prices and jobs today	25	40	31	41
People worry too much about human progress harming the environment	23	31	25	29
In order to protect the environment, the country needs economic growth	43	48	40	44
Economic growth always harms the environment	20	20	46	39
It is just too difficult for someone like me to do much about the environment	12	24	28	38

Source: Clarke and Stewart 1997.

Table 7.3a Polls on government action

	If you had to choose, which one of the following would be closest to your views?		
	Government should let ordinary people decide for themselves how to protect the environment, even if it means they don't always do the right things	Government should pass laws to make ordinary people protect the environment, even if it interferes with people's rights to make their own decisions	Can't choose
Canada	16	71	13
United States	17	62	21
West Germany	16	83	1
East Germany	10	89	1

Source: Clarke and Stewart 1997.

Table 7.3b Polls on government action

	If you had to choose, which one of the following would be closest to your views?		
	Government should let businesses decide for themselves how to protect the environment, even if it means they don't always do the right things	Government should pass laws to make businesses protect the environment, even if it interferes with businesses' rights to make their own decisions	Can't choose
Canada	5	88	7
United States	8	79	14
West Germany	9	91	1
East Germany	6	93	1

Source: Clarke and Stewart 1997.

Table 7.4 Polls on assessment of risk

Question asked	Percentage who replied 'extremely or very dangerous'			
(The 6 possible replies were 'extremely dangerous', 'very dangerous', 'somewhat dangerous', 'not very dangerous', 'not dangerous at all' and 'can't choose')	Canada	United States	West Germany	East Germany
In general, do you think air pollution caused by cars is ... for the environs?	60	47	62	57
In general, do you think air pollution caused by cars is ... for you and your family?	48	36	41	41
In general, do you think nuclear power stations are ... for the environs?	49	40	63	62
In general, do you think nuclear power stations are ... for you and your family?	46	35	55	46
In general, do you think air pollution caused by industry is ... for the environs?	78	61	79	76
In general, do you think air pollution caused by industry is ... for you and your family?	68	51	63	58
In general, do you think that pollution of the country's rivers, lakes and streams is ... for the environs?	74	66	73	80
In general, do you think that pollution of the country's rivers, lakes and streams is ... for you and your family?	65	53	54	58
In general, do you think that a rise in the world's temperature caused by the greenhouse effect is ... for the environs?	57	41	77	75
In general, do you think that a rise in the world's temperature caused by the greenhouse effect is ... for you and your family?	50	37	65	60
In general, do you think that pesticides and chemicals used in farming are ... for the environs?	52	38	64	52
In general, do you think that pesticides and chemicals used in farming are ... for you and your family?	45	35	53	41

Source: Clarke and Stewart 1997.

Table 7.5 Polls on willingness to pay higher prices and taxes

Question asked	Percentage who replied 'very willing or fairly willing'			
(The 6 possible replies were very willing, fairly willing, neither willing nor unwilling, fairly unwilling, very unwilling and can't choose)	Canada	United States	West Germany	East Germany
How willing would you be to pay much higher prices in order to protect the environment?	54	49	46	28
How willing would you be to pay much higher taxes in order to protect the environment?	37	38	33	19
How willing would you be to accept cuts in your standard of living on order to protect the environment?	45	32	53	41

Source: Clarke and Stewart 1997.

Nothing about the survey results suggests that Canadians cared less about the environment than Americans or Germans or that they were less willing to pay for environmental protection. The results also indicate that, compared to Americans, Canadians were less hesitant about government regulation for environmental quality. These results cannot explain why environmental protection has historically been weaker in Canada than the United States. These survey results make Canada's lackluster record in environmental protection all the more puzzling.

The argument here is that Canada's poor environmental performance is best explained by its highly decentralized approach to environmental policy. What follows is a description of the how responsibility for environmental policy is assigned in the Canadian federation. The section goes on to describe the limited role the federal government plays in Canadian environmental policy.

Canadian Environmental Policy and the Division of Powers

Canada is, in many respects, a very decentralized federation and particularly so in environmental policy. This noncentralization is a legacy of the 1867 constitution but it reflects political constraints as much as constitutional ones. Canada's federal arrangement presents several barriers to a strong national environmental policy: the constitutional division of powers, Quebec's separatist leanings, the desire of all provinces to maintain their autonomy and consensus norms in cooperation.

Constitutional questions cast a long shadow over all aspects of Canadian politics, directly or indirectly. Most areas of policy are fraught with conflicts over which level of government has jurisdiction, or conflict over how shared jurisdiction is to be exercised. In addition, the unresolved situation of the Canadian Constitution, which has lasted for almost 40 years, affects all aspects of government. The spectre of Quebec separation imposes constraints on policymaking. Not giving offense to provincial governments becomes a major objective of the federal government.

Table 7.6 Heads of power under the Canadian constitution relating to environmental policy

Heads of legislative power and areas of jurisdiction	Relevant Section of Constitution	Whose jurisdiction?	
		Federal	Provincial
Residual powers not enumerated in s.91 and s.92		x	
Federal lands		x	
Treaty making	originally s.132 of *BNA*	x	
'Extra-provincial works and undertakings'	s. 92 (10) (a)	x	
'Works for the general advantage of Canada'	s. 92 (10) (c)	x	
International trade and commerce	s.91 (2)	x	
Interprovincial trade and commerce	s. 91 (3)	x	
Spending power		x	
Residual power to make laws for 'peace, order and good government'	s. 91 (29)	x	
Fisheries (Sea coast and inland)	s. 91 (12)	x	
Agriculture		x	x
Direct taxation	s. 92 (2) s. 91 (3)	x	x
Indirect taxation of natural resources			x
Nuclear energy		x	
Local works and undertakings	s. 92 (10)		x
Municipal government			x
All matters of a merely local or private nature	s. 92 (16)		x
Property and civil rights	s. 92 (13)		x
Forestry	s. 92 (5)		x
Mining	s. 109		x

Source: Lucas 1987, 33.

In theory, environmental policy is an area of shared jurisdiction, but it has become an area of *de facto* provincial jurisdiction (Lucas 1986, 34). Most questions over division of constitutional power are resolved by reference to the *British North America Act* of 1867 and its successor the *Constitution Act, 1982*. The constitution of 1867 divided power between the federal and provincial governments in sections 91 and 92, which 'confer exclusive jurisdiction to pass laws in relation to the enumerated subjects' (see Table 7.6). In 1867, the Fathers of Confederation neglected to provide for matters such as telecommunications and environmental protection. Powers over these matters are derived from powers over analogous matters in the 1867 constitution. Thus, provinces have jurisdiction over environmental policy primarily by virtue of their role as the owners of natural resources in the provinces.

The Canadian federation has undergone several cycles of centralization and decentralization since its inception. As in the US, the Supreme Court has at times favoured federal government authority, at other times provincial authority. The Great Depression and World War II ushered in a period of centralization. The dramatic expansion of the welfare state in the 1950s and 1960s also saw an increase in the federal role. This expanded federal role did not extend to the environment, however. The environment became most politically salient (late 1960s, early 1970s) at the same time as Quebec's demands for greater autonomy, a political force which has barred further growth in federal powers and particularly, federal encroachment on provincial areas of jurisdiction.[7] Separatist sentiments in Quebec since the 1960s have sent the constitutional pendulum on a decentralizing trajectory from which it has yet to rebound.

The Federal Government:
Constitutionally Constrained or Self–restrained?

The federal government's limited role cannot solely be explained by the impasse over constitutional reform, however. Under the current constitution, the federal government has the legal authority to intervene more strongly in environmental policymaking than it actually has done. It has exclusive jurisdiction over fisheries, inland and coastal. While the federal *Fisheries Act* (s. 36) prohibits the disposal of 'deleterious substances' into waters that contain fish, the federal government has often been hesitant to use and enforce the *Fisheries Act*.

While the federal government has passed legislation governing fisheries, it has other powers which it has not used. In theory, the federal government could use its powers over criminal law, powers over taxation or its residual powers. In 1997, the Supreme Court upheld federal penalties under the *Canadian Environmental Protection Act (CEPA)* against a Quebec government utility on the basis of the

---

7   The first of several often fruitless constitutional conferences took place in 1971.

federal government's criminal law powers.[8] It is probably within the federal government's jurisdiction to introduce taxes on pollutants, such sulphur dioxide or carbon. The primary obstacles are political. Pollution taxes are difficult to introduce in any jurisdiction. To date, the Canadian federal government has never tested the extent of its powers of criminal law or taxation in environmental policy.

Of the federal government's residual powers, the most important is Peace, Order and Good Government (POGG). Prior to 1988, it was unclear whether POGG was a valid basis for federal environmental legislation. The Supreme Court's 1988 *Regina v. Crown Zellerbach* decision established that the federal government had jurisdiction over pollution in coastal waters, on the basis of POGG. The judges stated that use of POGG is justified on the basis of a doctrine of 'national concern'(Harrison 1996, 46). The judges specified that, for federal powers under POGG to apply, 'subject matters must be distinct, indivisible matters with a limited scale of impact on established provincial subject matters'(Lucas 1990, 29).

In turn, one of the tests of 'distinctness' is provincial inability. On this basis, the Supreme Court found that the federal government had authority to regulate 'persistent toxic substances' because their control is *distinct* from 'general waste control.' An expert in Canadian environmental law stated that, in the wake of this landmark decision, 'large chunks of the "environment" subject now appear to be fair game for federal legislators' (Lucas 1989, 183).

In the event of a direct conflict between a federal law and a provincial law, the doctrine of paramountcy means that the federal law 'takes precedence and the provincial law is inoperative to the extent of the conflict'. Generally, however, courts have taken a very restricted view of what constitutes direct conflict (Lucas 1987, 35). For political reasons, the federal government has chosen not to test the extent of its powers over environmental policy: '... the limitations on a strong federal role are not so much legal as they are political, with several provinces resistant to a greater federal role in environmental matters' (Muldoon and Valiante 1989, 26).

Further Evidence of Self-imposed Federal Restraint

Environmental policy is not the only policy area where the federal government has refrained from using its powers. The federal government has exclusive constitutional powers over interprovincial trade and commerce, similar to the Interstate Commerce Clause in the American Constitution. Whereas in the US, the Interstate Commerce Clause has been interpreted very broadly, authorizing extensive federal pre–emption in many areas of jurisdiction, the Canadian federal government has been remarkably hesitant to use these powers. Trade within Canada has been and continues to be hampered by a multitude of interprovincial trade barriers (Trebilcock et al. 1983).

---

8 *Canada v. Hydro-Quebec*. Thanks to Gord DiGiacomo for bringing this case to my attention.

The federal government clearly has the legal authority to pass legislation to deregulate interprovincial trade but it appears such legislation was never contemplated before 2007. In the 2007 Speech from the Throne, the federal government said it would 'consider how to use' its trade and commerce powers in the Constitution to 'make our economic union work better for Canadians' (Vieira 2007, FP1).

It is ironic that between 1987 and 1995, trade was, in theory, freer between Canada and the US than between Canadian provinces. The Canada US Trade Agreement introduced formal dispute settlement mechanisms for bilateral trade disputes; there was no comparable mechanism for resolving such disputes within Canada. Dismantling of interprovincial trade barriers was to begin under the auspices of the Agreement on Internal Trade (AIT), which entered into force in 1995. The AIT was subject to the kind of negotiation seen in bargaining rounds of the General Agreement on Tariffs and Trade: unanimous consent was required (Doern and MacDonald 1999).

When the AIT was negotiated, its lack of enforcement powers was identified as a potential weakness (Brown 2002). Subsequently, it fell well short of its goals with regard to reducing interprovincial trade barriers and resolving interprovincial trade disputes (Knox 2000). The determinations of the AIT dispute settlement process were often ignored by the guilty parties. In 2008, financial penalties for noncompliance were announced for disputes brought after 2009 (Council of the Federation 2008).

While the Canadian federal government's powers are circumscribed by the Canadian constitution in several areas of jurisdiction, its failure to intervene on the environment cannot be explained solely by constitutional powers. The Canadian federal government has chosen not to act, even when the constitution clearly grants it exclusive jurisdiction (such as for interprovincial trade), in deference to the very powerful provincial governments. Thus it is not surprising that the federal government is even more reluctant to take measures on the environment, where its constitutional bases for acting, while real, are more limited.

The Limited Federal Role in Environmental Legislation and Regulation

Compared to the US federal government or the European Commission, the Canadian federal government plays a very limited role in environmental legislation and regulation. The United States Environmental Protection Agency enforces fifteen pieces of legislation and hundreds of regulations. Prior to 1988, the only federal pieces of pollution control legislation in Canada were the *Fisheries Act, Canada Water Act (1970), Clean Air Act (1971), Ocean Dumping Control Act (1975)* and the *Environmental Contaminants Act (1976).* In 1988, all but the *Fisheries Act* were replaced by *the Canadian Environmental Protection Act.* As of 1998, there were 32 federal regulations to be enforced under these two pieces of legislation.[9] This is a very small number of regulations, for all forms of pollution. For example, by 1996, the province of Ontario had 300 standards for air pollution alone (Ontario 2000).

Over time, a pattern has emerged in Canadian environmental legislation. When the environment is a highly salient issue in public opinion, the federal government proposes initiatives and sometimes, legislation. These initiatives are then parried by the provinces, either by the introduction of their own legislation or by negotiating a retreat by the federal government. Most Canadian environmental groups have consistently sought stronger federal action and oppose greater discretion for the provinces (See Clark and Winfield 1996).

Early Demands for Federal Action

Demands for federal action on pollution have recurred for over a century. The late nineteenth century saw extensive debates over federal legislation to limit the dumping of sawmill waste into Canadian rivers, with federal legislation being passed in 1873, on the basis of federal jurisdiction over navigable waters (Benidickson 2007, 44). Canada has no national standards on sewage treatment but, as early as 1910, a Canadian Senator called for federal legislation in this area (Benidickson 2007, 177–182 *passim*). Speaking before the Senate, Senator Napoleon Belcourt pleaded that 'our noble rivers shall no longer be made the receptacles of the raw sewage of the country'. He had only to look behind the Canadian Parliament Buildings to see the heavily polluted Ottawa River, the city of Ottawa's main source of drinking water. Ottawa typhoid epidemics in 1911 and 1912 resulted in 174 deaths. The Senator pointed to the benefits of stringent measures introduced in Europe. In 1911, he introduced draft legislation to prohibit contaminating navigable waters but it did not pass. He continued to push for this legislation until 1915.

Numerous objections to his legislation were raised. Coastal communities saw no reason to treat sewage that flowed into the undrinkable oceans. The constitutionality of federal legislation was also raised as an issue. The legislators decided they should wait to act until the International Joint Commission (IJC) issued its report on boundary waters shared with the United States, a report which was four years late in appearing (1919). That year, the Canadian and American governments agreed that the IJC should draft a treaty or concurrent legislation to address the pollution of boundary waters. However, these binational efforts had collapsed completely by 1929.

The federal government paid relatively little attention to environmental issues prior to the 1960s. There was an isolated episode during the Second World War, when the federal government forced the Greater Vancouver Water District to begin chlorinating its water. The federal government had threatened to take over

---

9   Canada. House of Commons. Standing Committee on Environment and Sustainable Development 1998.

Vancouver's water system itself, using its powers under the *War Measures Act* (Hill 2006, 77).

Federal Action in the 1970s

Both the federal and provincial governments enacted environmental legislation in the 1970s. The federal government's work on the *Canada Water* Act preceded the peak of public concern on pollution by several years: the Department of Justice began drafting the bill in 1967 (Parlour 1981, 34). At that point, newspaper coverage of water pollution issues had increased dramatically from 1964, but was only 30 per cent of the peak of coverage that was reached in 1970 (Parlour and Schatzow 1978, 10). In the early 1970s, it appeared the federal government would play a substantial role in environmental policy, particularly for water, through its jurisdiction over fisheries. A legal scholar observed:

> When the [federal] *Canada Water Act* was introduced in 1970 it was in many ways 'milestone' legislation. The *Act* authorized the federal government to, among other things, enter into co–operative federal–provincial water quality management agreements, establish 'water quality management areas' where 'regional management agencies' could levy effluent charges to water users or prosecute persons who deposited unauthorized wastes. Ultimately, in the case of interjurisdictional waters, the federal government was authorized to act unilaterally to regulate water quality. In fact, at the time of the final revision of this paper (1988) no water quality management areas have been designated and no regional management agencies have been set up (Webb 1988, 18).

Thus, while the *Canada Water Act* implied that the federal government was going to exercise initiative in this area, very limited action resulted: 'the water quality provisions of the CWA ... simply were never implemented' (Harrison 1995, 422). No management agencies have ever been set up nor were effluent charges implemented.

No province had environmental legislation prior to 1970, although they had common law precedents on pollution and some had public health legislation. In 1970, some provinces had no water pollution control legislation, other than the federal *Fisheries Act* (Sinclair 1990, 99). The passage of the *Canada Water Act* in 1970 seems to have prompted water pollution legislation in at least two provinces. Alastair Lucas, an authority on Canadian environmental law, suggested that the provinces which passed legislation in the early 1970s (following passage of the federal *Canada Water Act* and *Clean Air Act*) did so primarily in order to prevent federal encroachment. He added that, in the years following, 'federal–provincial disputes over natural resources and energy and negotiations leading up to the patriation of the Constitution in 1982 had the effect of augmenting *de facto* provincial powers'(Lucas 1986, 34).

Also in 1970, the federal government amended *The Fisheries Act* to be able to set minimum national effluent standards, based on the abatement technology available at that time (Sinclair 1990, 90). The federal government promulgated several industrial effluent standards under *The Fisheries Act* during the 1970s. In discussing proposed Pulp and Paper Effluent Regulations in 1970, the Minister responsible said that national standards were needed to eliminate the possibility of 'pollution havens'.[10] Industry officials, such as the Council of Forest Industries countered that pollution control should be left to local or provincial authorities (Sinclair 1990, 99). Although regulations covering twenty industry sectors had been planned in 1977, only six regulations were ever issued (Harrison 1996, 100).

By the late 1970s, the federal government stopped enforcing its own regulations, leaving this task to the provinces. A 1980 study of environmental regulation across several industries found that:

> [i]n the face of an aggressive provincial stance and uncertainty about the extent of constitutional powers, federal authorities have been willing to yield the lead role in most environmental issues to the provinces. Case study authors ... describe the federal role in water and air quality management as limited to establishing minimum national effluent guidelines and standards so as to avoid the establishment of pollution havens in Canada. Compliance programs for existing sources of emissions in the pulp and paper and non-ferrous smelting industries are administered by the provinces, which also handle enforcement even of federal regulations (Thompson 1980, 22).

More stringent federal standards for industries did not take precedence over less stringent provincial standards, because of grandfathering of existing facilities (Sinclair 1988, 88).

In the *Clean Air Act* of 1971, the federal government pursued a two track strategy. First, the federal government gained the authority to set *National Emissions Standards* (minimum national emissions standards for particular industries) 'where necessary to prevent a significant danger to human health or a violation of an international air quality agreement' (Muldoon and Valiante 1989, 35–6). Only four standards were ever set however, all between 1978–1979. These four standards were later incorporated in the *Canadian Environmental Protection Act*.

Second, the Act also marked an attempt by the federal government to spur uniform action by the provinces. The Act provided for two kinds of nonbinding standards. There were ambient National Air Quality Objectives, negotiated jointly with the provinces, which were supposed to be adopted into provincial legislation. In addition, the Act provided for unenforceable *National Emission Guidelines* 'intended to encourage uniform standards across the country through their

---

10   Jack Davis, Minister of Fisheries and Forestry, quoted in Sinclair (1990, 92).

adoption by the provinces' (Muldoon and Valiante 1989, 35–6). Between 1974 and 1981, guidelines were set for seven industry sectors. No studies have examined the extent to which these guidelines were adopted in provincial legislation.

Federal Action during the 1980s

In the late 1980s, Canadian public opinion polls detected a sharp increase in the salience of environmental issues. The federal government launched two initiatives in this period: regulations under the new *Canadian Environmental Protection Act (1988)* and *Canada's Green Plan*. A *Drinking Water Safety* Act was announced but it never became law. It would have had a very limited ambit, covering only places under federal jurisdiction, such as Indian reservations, military bases and airports (Hill 2006, 62).

In a continuation of the non–coercive approach, the federal government unveiled *Canada's Green Plan*, its guide for implementing the 1987 Brundtland Commission's concept of 'sustainable development'. In the *Green Plan*, the federal government tried to use the spending power to affect environmental policy in Canada, which it had previously used to greatly expand its role in healthcare and welfare. Although there was a brief flurry of use of the spending power with the announcement of the federal *Green Plan* in 1990, it was not sustained. The *Green Plan* promised to spend over C$1 billion on environmental protection over five years. However, little of this money represented new expenditures: most of the dollars were just renamed or shuffled around. After a few years, the *Green Plan* was quietly allowed to die before its five years were up.

Federal Promotion of Provincial Environmental Policy

The federal government has sought to strengthen Canada's national environmental policy by stimulating provincial action and entering into negotiated agreements with provincial governments. The first initiative was launched in 1975, when the federal government announced that it would negotiate accords with the provinces to delegate responsibility for environmental regulations. The second was the advocacy strategy of the early 1980s. The third was the passage of federal legislation to regulate toxic substances, the *Canadian Environmental Protection Act* (CEPA) in 1988. All these initiatives have been marked by a federal retreat from enforcement of its own regulations.

The Federal–provincial Accords of 1975

The year 1975 marked the federal government's first formal attempt at pushing the provinces to address the environment. Federal agreements known as 'Accords for the Protection and Enhancement of Environmental Quality' were signed with seven provinces (not Quebec, British Columbia or Newfoundland), for five years duration. Only Alberta and Ontario subsequently renewed for another 5 years. One section of the accords pointed to the Holy Grail of the federal government's cooperative efforts: *binding minimum national standards arising from federal–provincial negotiation.* Section 9 of the Accords stated: 'Canada, after consultation with the Province and all other Provinces, agrees to develop national baseline effluent and emission requirements and guidelines for specific industrial groups and specific pollutants' (Giroux 1987, 86–7). Had this effort been successful, Canadian regulations might have more closely resembled the American regulatory framework of industry specific emissions standards for particular pollutants. Instead, Canadian regulations have tended to avoid setting binding emissions limits for specific industries and pollutants.

There is little to suggest that the Accords had any effect except a drastic reduction in the enforcement of existing federal regulations, in at least some jurisdictions. One legal scholar noted: '[a] number of legal features of the Accords are worth noting in order to draw some policy conclusions about their usefulness. Most important is the fact these Accords have no legal status'(Giroux 1987, 87). Another worrisome detail is that the federal government ceased to enforce its regulations in those provinces which had *not* entered into accords. In the case of Quebec, this resulted in the wholesale neglect of water quality because Quebec had few standards of its own. In assessing these developments, the Law Reform Commission of Canada concluded:

> [o]ur review of the Environmental Protection Service (EPS) delegation scheme suggests that federal delegation of environmental enforcement authority *amounts to a virtual abdication of responsibility* for enforcement and has promoted discrepancies in the nature of enforcement responses across Canada (Heustis 1984).

With regard to the obstacles to effective provincial environmental policy, excessive federal interference is clearly not one of them.

Federal Retreat: The Advocacy Strategy

By the early 1980s, the federal government officially rolled over and played dead when it introduced its 'advocacy strategy'. The federal government would no longer enforce any of its own environmental regulations, leaving this task entirely to the provinces. Henceforth the federal government would put its resources into research and moral suasion, not legislation or enforcement:

> [t]he new approach has also included the concept that departmental policies should try to influence and persuade rather than play a strictly repressive role. [The advocacy approach] gained such importance that concerns for implementing laws and regulations decreased from 1982 to 1984, with adverse repercussions on legal procedure initiatives (Giroux 1987, 86).

Delegating implementation of federal regulations on air and water quality control
led to wide discrepancies in enforcement (Giroux 1987, 86).

The Canadian Environmental Protection Act and Equivalency

The federal government's third effort to spur provincial action was part of the
Canadian Environmental Protection Act (CEPA), intended to provide a national
framework for addressing toxic substances. Minister of the Environment Thomas
McMillan announced the proposed *Environmental Protection Act* in 1986 with
these words:

> [a] good law, however, is not itself enough. It must be enforced – ruthlessly
> if need be. Accordingly, the new *Environmental Protection Act* will be
> accompanied by a plan to reverse the country's appalling record of enforcement
> and compliance.

It is striking for a Cabinet Minister to describe his country's record as 'appalling'.
While McMillan acknowledged the inadequacy of past efforts in Canada, the
legislation he introduced did little to alter the established course.

The federal government chose to pass legislation on toxic substances because
it seemed to have two good constitutional bases. First, as environmental lawyers
Paul Muldoon and Marcia Valiante argued (1989, 26): '[a] national toxic chemical
control strategy could be considered as falling within the "national concern"
test, owing to the possibility of environmental pollutants to traverse political
boundaries, the diverse sources, and the seriousness of the problem in most parts
of Canada'. Secondly, because of the federal government's trade and commerce
power, the manufacture, use, transportation and importation and exportation of
toxic chemicals have been accepted as being within federal legislative authority,
whereas disposal of toxic wastes are under provincial jurisdiction (Muldoon and
Valiante 1989, 26).

Despite this apparently strong constitutional basis, when the federal government
passed regulations on dioxin from pulp production, the regulation was produced
by interprovincial negotiation. Under *CEPA*, dioxins were subject to a goal of
'virtual elimination'.[11] As a result of a high level of attention from the public
and Greenpeace, the Canadian industry did not offer much resistance to new
regulations (Marotte 1990, 7). In a detailed study of how dioxin regulations were
set for the Canadian pulp and paper sector, Kathryn Harrison found that the federal
government acquiesced to the provinces wanting the least stringent standard,
including Quebec. Alberta, another province with anti–federal proclivities, had

---

11   In 1988, a landmark study by the US EPA confirmed that pulp mills using chlorine
bleaching were discharging significant amounts of dioxin in their effluent. This led to rapid
action on dioxin in pulp mill effluent by major pulp producing countries like Sweden and
the US.

favoured a higher standard. Quebec then unilaterally set its standard slightly higher than the federal standard, to prove it had no need of federal standards (Harrison 1996). Once again, the goal of a minimum national standard was thwarted.

Keeping provincial sensibilities in mind, *CEPA* as a whole was subject to the principle of 'equivalency'. Negotiations over equivalency were intended to establish the extent to which this new legislation would complement or supplant existing provincial powers. If provincial regulations found to be 'equivalent' to those proposed by the Canadian government under *CEPA*, they would take precedence. The concept was soon put to the test after a major fire in a Polychlorinated Biphenyls (PCB) storage facility in St. Basil-le-Grand in Quebec:

> The Mulroney government, determined to appear capable of decisive action during a federal election, immediately announced that it would use its newfound powers under *CEPA* to regulate PCB storage directly throughout the country, upon which all of the provinces except Prince Edward Island, equally determined to resist any expansion of federal power, applied for exemption from such federal regulation on the basis of equivalency (McInnes 1988, A12).

Only 15 of Canada's 2000 PCB storage sites were located in the province of PEI, therefore the vast majority of sites would have been exempted from federal legislation. By 1994, Alberta had concluded an equivalency agreement with the federal government, covering four regulations from the revised *CEPA* of 1999. As of 2008, this was the only equivalency agreement ever negotiated (Environment Canada, CEPA Environmental Registry).

In the late 1990s, two reports to Parliament assessed the federal government's practices for enforcement of federal regulations. A 1998 report by the House of Commons Standing Committee on Environment and Sustainable Development examined the enforcement under the *Fisheries Act* and the *Canadian Environmental Protection Act*. It stated that

> [t]he Committee is at a loss to understand why [Environment Canada] does not have comprehensive, standardized and readily accessible data on enforcement budgets and expenditures. Without this basic information, it is difficult to see how the Department can evaluate its enforcement program and determine whether or not changes have to be made (Canada. House of Commons. Standing Committee on Environment and Sustainable Development 1998).

With regard to bilateral agreements with the provinces, several witnesses testified that their enforcement was 'extremely problematic', in part due to the lack of publicly available information. The report stated:

> ... the fact remains that Environment Canada and indeed some provinces are not enforcing environmental laws when they could and should. This failure to act is of deep concern to the Committee. It is doubly troubling in light of the

federal government's decision to enter into a larger harmonization agreement covering the entire country. In the Committee's opinion, it is essential that the administrative agreements establish rigorous management structures and accountability (Canada. House of Commons. Standing Committee on Environment and Sustainable Development 1998).

The report recommended that the Auditor General of Canada conduct an effectiveness audit of all bilateral federal–provincial agreements which delegate enforcement to the provinces.

Acting on that recommendation, the Auditor General of Canada reported on seven bilateral federal–provincial enforcement agreements in 1999. The report found that the Parliament of Canada was 'receiving incomplete and outdated information on the results of the *CEPA* agreements, and no information on the results of the *Fisheries Act* agreements' (Office of the Auditor General 1999). The report found that several agreements had not been properly implemented and noted that the federal government had no plan for resumption of its enforcement responsibilities, in the event a province were unable to carry out its assigned responsibilities.

The Auditor General's report also noted that, while all the agreements focused on streamlining administration, only two mentioned environmental protection as an explicit objective. For the other five agreements without environmental protection as a stated objective:

> Environment Canada informed us that it does not think it is possible to evaluate the extent to which the agreements have contributed to measured environmental improvements.

On the whole, the report found that the federal government was not doing enough to ensure that its enforcement responsibilities continued to be carried out. The *CEPA* was revised in 1999, but many observers contend the changes did nothing to strengthen enforcement (Baxter 1999, A5; Jacobs 1999, A1).

The record shows that the Canadian federal government has played a circumscribed role in environmental policy, particularly in pollution control. Federal regulations of industrial pollutants are conspicuously few. Furthermore, most federal incursions into environmental policy have, over time, been eroded by delegation to the provinces. In more centralized federations, a description of the federal regulatory framework provides a fairly complete picture of the national framework. In Canada, the federal and the national are clearly not interchangeable. Getting a sense of the national picture requires examining the patchwork quilt of the policies of thirteen provincial and territorial governments.

The Canadian Regulatory Patchwork

In face of such a limited federal role, what does Canada's regulatory system look like? How well does it perform? Unfortunately, if information about the federal regulations is difficult to obtain, it is even more difficult to form a national picture from the patchwork of provincial regulations. It is possible to form a partial picture from three regulatory domains where more information has been available: the pulp and paper sector, drinking water regulation and municipal waste water treatment.

Regulating Industrial Pollution: The Canadian Pulp and Paper Industry

There is remarkably little written about environmental regulation of Canadian industry and its effect on the quality of the environment. Lack of concrete data, poor access to information and the overall fragmentation of environmental policy contribute to the problem. Only one Canadian industry's regulatory framework and environmental performance has received systematic study over time: the pulp and paper industry. This is the only industry for which there is has been a publicly available study (1990) comparing federal and provincial regulations, comparing those to standards in other countries and presenting data on compliance with those standards (Sinclair 1990).

In the early 1980s, two fairly comprehensive studies of the Canadian regulatory system appeared, although they presented little data on enforcement and no information on environmental quality outcomes (Franson, R. et al. 1982; Thompson 1980). Both reports covered the pulp and paper industry. Both found that negotiation between government and industry was an important feature of the Canadian regulatory process at all stages, from standard setting to enforcement. Another study confirmed this finding for the pulp and paper sector in British Columbia (Nemetz 1986, 405). An economic study from this period found that, on a pro rated basis, Canadian industries spent approximately *one–third* of what their American counterparts did on environmental protection. The report attributed the difference solely to more stringent regulations in the US (Economic Council of Canada 1981, 88).

In 1988, an Environment Canada report on the implementation of the *1971 Fisheries Act* regulations was leaked to the press (Bohn 1989, A1).[12] The report presented data on the discharge limits which applied to each of the 122 pulp mills in Canada in 1985, as well as data on whether the facility's effluent met the federal standard in 1985. (The report (Sinclair) was eventually published in 1990.) The headline on the front page of the *Vancouver Sun* was somewhat misleading: 'Study Reveals That Majority of Pulp Mills Break Law: Worst Polluters in Canada Found in BC and Quebec' (Bohn 1989, A1). A majority of mills did not meet the federal

---

12   The 1971 regulations covered only Biological Oxygen Demand (BOD), Total Suspended Solids (TSS) and Acute Toxicity to Fish.

standard, but they were not legally obligated to do so. An official spokesman from the Council of Forest Industries of BC pointed out that most mills in BC were subject to provincial, not federal standards. He stated, '[i]t just so happens that those standards are less strict than the federal ones', adding that BC toxic chemical limits for coastal mills were substantially less strict than the federal standard under the *Fisheries Act* (Bohn 1989, A1).

The explanation for this discrepancy is the common practice of grandfathering new regulations. The 1971 *Fisheries Act* amendments were legally mandatory only for facilities that were new or renovated after 1971. For mills built prior to 1971, the federal regulations were merely guidelines and the mills were subject to provincial discharge limits only. In 1986, only eleven of Canada's 122 mills were 'new' and therefore completely subject to the new federal rules. (An additional forty mills had been expanded and the expanded portion of the mills were subject to the federal regulations).

There was an assumption that, over time, all mills would be brought under the federal standard and that the regulations would serve as a target for existing mills (Sinclair 1990, 111). Fourteen years after the passage of the federal regulations, a substantial percentage of mills across the country did not meet the federal standards, despite a significant subsidy program for pulp mill modernization (Sinclair 1991, 90).

Using 1985 data, the study by Sinclair compared the discharge limits in each mill's permit to the *Fisheries Act* regulations (without regard to whether or not that mill was legally required to comply with the regulations). For approximately 7 per cent of mills (all but one in Ontario), no information on discharge limits was available. Forty–five per cent of Canadian mills had permits with TSS limits more stringent than the federal standard. However, an additional 31 per cent of mills were subject to TSS limits less stringent than the federal *Fisheries Act* standard, including 96 per cent of mills in British Columbia (Sinclair 1990, 88). Similarly, 45 per cent of mills were subject to BOD discharge limits more stringent than the federal standard. But 31 per cent were subject to BOD limits less stringent than the federal standard, including 58 per cent of mills in Quebec (Sinclair 1990, 89).

The compliance record of those mills subject to federal regulations was not particularly good. Looking at the new or expanded mills that were at least partially subject to the federal regulations, Sinclair found that only one third of these were in compliance with the federal toxicity standard as of 1985 (Sinclair 1990, 187). Of all 122 mills, 68 per cent did not meet the federal toxicity standard for toxicity to fish. Of all the mills, 39 per cent were not in compliance with the federal TSS limits and 21 per cent did not meet the federal BOD limit. Sinclair argued that, while most mills could achieve compliance with the TSS and BOD standards with minor reductions, the same could not be said for acute toxicity. Acute toxicity posed the greatest long term threat to the environment but was also the most challenging to control, unfortunately (Sinclair 1990, 245).

In theory, provincial governments could have used their standard setting authority to tailor discharge permits to the condition of receiving waters. This is what advocates of decentralization say will happen. Because the federal standard was based on 'best practicable technology', it did not take into account variations in the adaptive capacity of receiving waters. Discharging pollution into shallow waters will have a greater environmental impact than the same amount of pollution into deeper water.

Sinclair's report did not find evidence that provinces tailored their standards by applying more stringent standards in more vulnerable receiving waters (Sinclair 1991, 93). This finding is confirmed by a study of the impacts of pulp mill effluents on Canada's east and west coasts: '... effluents from Canadian coastal mills were usually acutely toxic at source and in many cases had marked deleterious effects due to toxicity, high BOD and TSS loadings ... Integrated site specific assessments need to be undertaken to document ecosystem response to process and treatment improvements at mill sites' (Colodey and Wells 1992, 201).

Sinclair's report was unique in comparing Canadian environmental regulations for an industry to those of other countries. It indicated that, as of 1985, US EPA BOD limits for pulp and paper mills were more stringent than federal or Ontario limits for *any* of the 20 Ontario mills for which information was available. If US EPA limits for TSS had applied to Ontario mills, these would have been more stringent than federal or provincial limits for 77 per cent of mills (Sinclair 1990, 222). Because Ontario's limits were at least as stringent as those in other provinces, the rates of compliance in other provinces would have been even lower. In examining pollution control policy for the sector in Sweden and Finland, Sinclair concluded that the control policies of these governments were more aggressive in pushing pollution reduction than the Canadian policies (Sinclair 1990, 225–6). Thus, for one of Canada's most heavily polluting industries, federal and provincial emissions standards were significantly less stringent than comparable standards in other countries.

Sinclair concluded that,

> the negotiation pattern that has been established benefits those who wish to delay making investments in pollution control and works to the disadvantage of government officials responsible for controlling pollution from the industry ... Unfortunately, concessions made to individual mills on the basis of arguments of questionable economic merit further discourages other mills from willingly investing in the necessary controls (Sinclair 1990, 260–1).

Kathryn Harrison, studying the same sector in the 1990s, found that the negotiated Canadian approach led to significantly lower rates of compliance in comparison with the US industry (Harrison 1995). A study of the pulp and paper industry in Quebec found that both inspections and the *threat of an inspection* had strong negative impacts on emissions and that inspections also induced more frequent self–reporting from industry (Laplante and Rilstone 1996).

Having examined how the pulp and paper sector had been regulated over several decades, William Sinclair criticized the tradeoffs made by the federal and provincial governments. He found that they had overestimated the negative economic impact of pollution abatement and overestimated the benefit of concessions in terms of employment and income (Sinclair 1991). He concluded:

> Government environmental authorities have overestimated the negative economic impact of mills adopting the technologies necessary to control effluents; they have overestimated the degree to which Canadians benefit from what are intended as temporary concessions on discharge controls to protect employment and income. Perhaps even more important, environment authorities have failed to take into account adequately the enormous costs they may be imposing on present and future generations of Canadians ... (1991, 86).

Governments' fears of job loss or investment forgone are a substantial obstacle to better environmental policy in Canada. Faced with the threat of factory closure, Canadian governments have usually caved in to industry demands. This is particularly true in one–factory towns in outlying regions, which have higher unemployment than cities (MacDonald 1991, 226). In the 1980s, 50 per cent of the pulp and paper industry's capacity was located in communities of fewer than 10,000 people (Sinclair 1990, 288). However, many of these threats of exit or closure were not credible. For example, 21 per cent of the pulp mills in Ontario and Quebec in 1987 had begun operating in the nineteenth century, suggesting that the industry infrequently closed its factories (Sinclair 1990, 72).

What general conclusions can be drawn from this overview of the regulation of the Canadian pulp and paper sector? First, it offers no support for the notion that Canada has been a leader in controlling pollution from this very highly polluting sector. It also conclusively demonstrates that, while some provinces will, on occasion, exceed minimum federal standards, the provincial requirements for many mills fell well short of the minimum federal standard, even fourteen years after these federal standards had been introduced. Thus there is no evidence for a provincial race to the top in environmental standards for this sector. Nor is there any support for hypotheses that subnational governments with significant autonomy will tailor pollution control measures to local conditions and that they will use that autonomy to innovate. There is no evidence for provincial learning or innovation here.

Further, the negotiated approach to permitting and compliance was not conducive to high levels of environmental protection or to technical progress. Because permit conditions were negotiated and enforcement over compliance was also negotiated, this introduces several points of slippage in the standard setting process. There is downward pressure on standards during the negotiating of the permit. (As Harrison noted, there is no evidence that negotiation at the permitting stage 'buys' better compliance, obviating the need for enforcement.) This is followed by further laxity introduced by poor enforcement. All of the studies cited above support the argument that consistent environmental protection requires consistent enforcement of binding regulations.

Regulating Publicly Owned Sources: Drinking Water and Wastewater

It might be argued that Canada's weak performance in regulating industrial pollutants is a product of its economic structure. Historically, the Canadian economy has been based on extracting and refining resources. If this explained the overall pattern of environmental protection, we should expect to see regulations and performance comparable to other developed countries in publicly owned sectors that are more insulated from economic competition. This is not the case, however. In environmental regulation pertaining to urban services, Canada also performs poorly in comparison with the United States and the European Union. The analysis here focuses on standards for drinking water and municipal wastewater treatment.

Drinking water quality   Whereas the United States and the European Union have set mandatory minimum standards for drinking water, Canada has no national regulations in this area. The American regulation covers 80 parameters. Since 2003, the EU regulation covers 45 parameters. The first Canadian guidelines for drinking water standards were issued in 1968 and were revised in 1978. In 1986, a Federal–Provincial–Territorial Subcommittee on Drinking Water Quality was established to revise the voluntary guidelines in light of new scientific information, subject to the agreement of all provinces. The Committee (with 14 voting members), makes decisions by consensus and meets twice a year.

In 2008, the Canadian Guidelines covered 113 parameters for drinking water quality: microbiological (7), chemical and physical (88) and radiological parameters (18). The federal Commissioner for the Environment and Sustainable Development reported in 2005 that, of the 83 guidelines for chemical and physical parameters, about fifty were at least 15 years old and in need of updating (Office of the Auditor General 2005, 2). Because the review of each standard takes five years on average, the Commissioner estimated it would take at least ten years to clear the backlog (Office of the Auditor General 2005, 11).

Despite the extensive involvement of all provinces in this standard setting process, only a minority of provinces have adopted the Canadian guidelines in legally binding provincial standards (Hill 2006, 63–4). In 1972, Alberta became the first to adopt them, incorporating the guidelines into its binding *Municipal Plant Regulation* of 1978. Quebec adopted its own binding standards in 1984, prior to formation of the Federal–Provincial Committee. In 2002, the government of Ontario replaced its non–binding provincial drinking water objectives with binding standards for drinking water quality. This change was made after polluted drinking water in Walkerton, Ontario sickened 2,300 people and killed seven (Hill and Harrison 2006, 241). Even where binding provincial standards have been in place, it is difficult to draw conclusions about enforcement: '[n]o province specifically reports prosecutions or convictions related to drinking water'.[13]

---

13   Canadian Environmental Defence Fund (2001) quoted in (Hill 2006, 64).

As of 2006, 38 years after the first guidelines were published, four of the 13
provinces and territories (Nova Scotia, Ontario, Quebec and Alberta) had adopted
most or all of the voluntary guidelines into their provincial standards (Hill et al.
2007, 388–9). There was significant variation across the other jurisdictions, even
though the cooperative standard setting process had been in place for 20 years
(Table 7.7).

**Table 7.7 Drinking water quality parameters monitored by the
provinces and territories**

Jurisdiction	Number of parameters monitored for in drinking water (2006)			
	Microbiological	Physical and chemical	Radiological	Total
Alberta	Fully adopts Canadian Guidelines for Canadian Drinking Water Quality			
Nova Scotia	Fully adopts Canadian Guidelines for Canadian Drinking Water Quality			
New Brunswick	According to plan agreed upon by Minister of Health			
Manitoba	To approval of Medical Health Officer			
Newfoundland and Labrador	3	28	0	31
Prince Edward Island	–	–	–	50
Quebec	4	60	–	77
Ontario	3	70	0[1]	73
Saskatchewan	4	61		65
British Columbia	3	Additional parameters at discretion of Drinking Water Officer		
Yukon	2	30	–	32
Northwest Territory	1	31	–	32
Nunavut	1	23	–	24

Source: Hill et al. 2007, 378–9.
Notes: [1] Ontario has 78 radiological parameters but does not require monitoring for any of
them.

A Canadian environmental group compared the 2006 Canadian Guidelines to
drinking water standards in the US, Australia and the European Union. In general,
the Canadian guidelines were less stringent than those found in other jurisdictions.
For example, the US EPA requires filtration of all public water systems, whereas
only five of the ten provinces do (David Suzuki Foundation 2006, 12). The

Canadian guidelines for turbidity, a microbiological parameter, were less stringent than those of the US, EU and Australia. The Canadian guideline for *Escherichia coli* is somewhat less stringent than those of the other jurisdictions. Comparing the Canadian guidelines to WHO recommendations and standards in the other three jurisdictions for 67 contaminants, the report found that Canadian guidelines were less stringent than at least one of those standards for 53 of those 67 parameters (David Suzuki Foundation 2006, 14).

It could be argued that standards for drinking water in Canada are superfluous, due to the purity of source waters, but this view is not supported by the evidence. In 1989, years before the Walkerton tragedy, 95 per cent of Canadians polled stated they were concerned about the quality of drinking water in Canada (Nichols and Jensen 1990, 30–1). And they were probably had reason to be. Comparing drinking water quality in Canada and the US, Carey Hill (2006) found that American national standards, introduced in 1974, regulated a larger number of substances and set more stringent standards than corresponding provincial regulations. In a detailed analysis of drinking water quality (not just standards), Hill compared the quality of water in paired Canadian and American cities. She found that Americans enjoyed consistently better drinking water across the country, with less variation across cities than Canadians did. Furthermore, she found that the American cities experienced less day to day fluctuation in drinking water quality than Canadian cities did.

Municipal wastewater treatment   As discussed above, many Canadian cities lag far behind their American and EU counterparts in sewage treatment. A few still have no sewage treatment. Few cities match the highest levels of protection found in other developed countries. Many of those with treatment systems do not meet the minimum standards of secondary treatment found in the US and the EU. Why is Canada's performance so poor in this area? In comparison with other federal systems in this book, Canada differs in two respects. First, it has no minimum national standard for municipal wastewater treatment and provincial standards show great variation. Second, Canada has provided few federal incentives for sewage treatment plant construction. In the US, Switzerland and the EU, central government provided sticks (minimum standards) and carrots (grants or loans for sewage treatment plant construction).

Both the US and the European Union legally require a minimum standard of secondary treatment of sewage. In the US, the *Federal Water Pollution Control Act* (*Clean Water Act*) of 1972 required that all existing sewage treatment plants provide a minimum of secondary treatment by 1977 (Patrick 1992, 64). The European Union's 1991 *Urban Waste Water Treatment Directive* (91/271/EEC), also requires that all cities (greater than 10,000 people) subject all their sewage to secondary treatment.

Again, it could be argued that minimum national standards would be superfluous, if provincial standards are equivalent. However, there is no evidence to suggest that, on the whole, provincial efforts are comparable to US or EU standards. In addition to a minimum standard of secondary treatment, the EU directive also specifies emissions limits for all sewage treatment plants, for the following parameters: Biological Oxygen Demand (BOD), Chemical Oxygen Demand (COD) and Total Suspended Solids (TSS). There are additional effluent standards for sewage treatment plants in sensitive areas: total phosphorus and total nitrogen.

A 2004 review by the Canadian Council of Ministers of the Environment (CCME) of provincial laws and regulations on sewage treatment plants found that the following parameters were 'monitored' at 'most' sewage treatment plants in the jurisdictions in question:

- BOD – all 13 provinces and territories
- COD – two provinces
- TSS – all 13 provinces and territories
- total phosphorus – nine provinces/territories
- nitrogen (not necessarily total) – eight provinces/territories.

Compared with the European Union's standard promulgated thirteen years earlier, Canadian provincial requirements in 2004 generally did not include COD.

Although the CCME report indicates these parameters were monitored at most sewage treatment plants, there is no information on the level of any standards, if any standards were mandatory or if these standards are enforced. Thus municipal wastewater treatment marks another domain where it is difficult to ascertain what standards apply and whether these are enforced. The available information does not suggest that the provincial regulations taken together offer a level of protection equivalent to European Union standards.

Efforts to create a national strategy in municipal wastewater took shape in the 1990s. A CCME working group on the issue was formed in 1995. In 2003, the CCME agreed to prepare a Canada–wide Strategy and a draft discussion document was released for public comment in late 2007. The draft strategy emphasizes information exchange and the need for flexibility. In 2007, the Canadian federal government announced that, in 2008, it would publish regulations on sewage treatment and also provide infrastructure funding for wastewater treatment plants (Environment Canada 2007b). (As of May 2009, no federal regulations had been promulgated.)

The paucity of federal funds for wastewater treatment is another feature of the Canadian response that distinguishes it from that of the US, Switzerland and even the EU. Despite the many millions of dollars the Canadian government has spent on regional development in the post–World War II period, the federal government had only one brief program for sewage treatment plant construction, prior to 2000. Between 1961 and 1972, a federal government agency made 2,347 loans totalling C$627.6 million for sewage treatment plant construction. The US federal government, in contrast, spent billions of dollars in grants, beginning during the Great Depression until the 1990s. In order to help newly acceded EU members implement the 1991 directive, the EU projected spending at least 500 million € per annum for environmental investments, between 2000–6, through the Instrument for Structural Policies for Pre–Accession (ISPA). The total costs of complying with the directive in the accession countries was estimated to be between 79 and €110 billion (European Commission. DG Environment 2001).

The Canadian trend was partially reversed in 2000, when the federal government announced the Infrastructure Canada Program. Between 2000 and 2007, the federal government spent $960 million on wastewater treatment infrastructure, which was matched by provincial and municipal spending for a total of $2.88 billion resulting from the new federal initiative (CCME 2007, 3). Given the size of the problem, these were relatively modest sums. For example, the city of Ottawa budgeted $200 million for water and wastewater infrastructure construction costs between 2003 and 2021 (City of Ottawa 2007). The 2004 OECD Performance Review estimated that, at the current rate, it would take another twenty years for Canada to build the necessary wastewater infrastructure. The report added that it would be a challenge to make the necessary investments while also improving drinking water infrastructure (OECD 2004a, 64).

The above overview reiterates the difficulty of forming a composite picture of environmental regulation in Canada, due to gaps in information and substantial variation across provinces. The available information does not paint a pretty picture. In comparison with other federations, Canada has relatively few binding regulations for pollution control. The regulations in place are not particularly stringent, in international comparison.

Furthermore, both regulations and non–binding guidelines are updated infrequently, which strongly suggests that they do not reflect the current state of scientific information on impacts or the most up to date abatement technology. Requiring consensus among 10 to 14 governments to establish even a guideline contributes to the rapid obsolescence of the standards set. It is unclear what the benefit of consensus decision–making is here, because only a minority of provincial and territorial jurisdictions go on to adopt (or exceed) these guidelines in their own regulations. As a form of governance environmental protection, this highly decentralized model emphasizing cooperation has very little to recommend it.

Drawing Conclusions About Decentralization from the Canadian System

It is clear from the above that the Canadian system of environmental regulation is a very decentralized one in comparison with the US, Switzerland and even the European Union. Thus the Canadian case is ideal for examining claims made for and against decentralization of environmental policy in a developed country context. These claims focus on information, regulatory competition, innovation, tailoring standards to local conditions and reflecting public preferences.

Claims About Information

The Canadian system has not performed well in generating environmental information, nor in making it publicly available. This has made it very difficult for anyone, including researchers, journalists and activists, to determine what trends or patterns characterize environmental protection in Canada. As noted, the OECD's 2004 *Review* found substantial gaps in Canadian environmental information. To the extent these have been filled, it has been as a result of federal initiatives such as the National Pollution Release Inventory. Provincial governments have not collected data or required its dissemination, even on matters of public concern such as drinking water quality.

If it has been difficult to measure environmental quality over time and across Canada, it has been even more difficult to determine what provincial regulations are in place, if these have been complied with and what, if any, enforcement actions have been taken. For example, there are no accessible databases that compare all pollution control regulations or industrial permits across the provinces, or compliance with those permits (Olewiler 2006, 125). Within provinces, regulatory regimes have historically not been transparent or open to public scrutiny: both environmental standard setting and their enforcement have been subject to bargaining between industry and the regulator, typically behind closed doors (Thompson 1980, 33).

Political scientist Kathryn Harrison presents a stark contrast with the American situation:

> [i]n seeking quantitative measures of enforcement activity, students of the US regulatory regime are at a distinct advantage. The open and formal US approach to regulation has produced centralized data bases of enforcement activity at both the state and federal level. Records are not only kept; they are also publicly available. In contrast, the closed and informal nature of Canadian enforcement has meant that consistent records often have not been kept and where they do exist, they can be difficult to access... [n]ationally consistent records of enforcement activity (for example, inspections, warning letters and prosecutions) simply do not exist (1995, 228–9).

Here Harrison was discussing the enforcement of federal regulations, usually by the provinces. It would be impossible to compare the enforcement of the provinces' *own* regulations, across provinces.

This lack of transparency is not limited to the regulation of industry. It is almost as difficult to determine the quality of local tap water and whether provincial drinking water standards are complied with or are being enforced. Comparing the Canadian and American regulatory regimes for drinking water, Carey Hill found that the Canadian system is characterized by 'hidden information'. She noted that those governments with the readiest access to information (municipal governments) had the *least incentive* to collect and disclose such information, because it might impose costs on them and result in a loss of public confidence (Hill 2006, 219). Hill found that, even after the Walkerton Ontario tragedy, in which seven people died from drinking contaminated water from a municipal system, only Ontario, Saskatchewan and Newfoundland required annual public reports from municipal utilities. Prior to Walkerton, no provinces had annual reporting requirements (Hill 2006, 228).

We observe a paradox: local systems do not give rise to useful information about local environmental policies or local environmental quality. Higher levels of government are instrumental in providing information about local environmental conditions. Lower levels of government have not provided local publics with information about local environmental quality or regulatory activities. In Canada, as in the US and EU, local information has become available as a result of national (or EU level) requirements for data collection and transparency. Thus, claims for decentralization that assume greater or better knowledge of environmental conditions at the local level (rather than the national level) are not supported by the Canadian case.

Regulatory Competition: Races to the Bottom? To the Top?

A major argument against decentralized environmental regulation is that it leads to regulatory competition, seen in the form of races to the bottom. Races to the bottom result in regulations which are too lax, relative to public preferences. The absence of comparable data over time and across provinces has made this claim difficult to assess in the Canadian context (Olewiler 2006, 113).

As part of a project on interprovincial competition, economist Nancy Olewiler examined selected provincial environmental policies for evidence of races to the bottom. Because of the difficulties in obtaining data, she examined only the six provinces with the largest populations.[14] Comparing ambient concentrations of major air pollutants across the provinces, Olewiler found that, other than for sulphur dioxide, concentrations were converging. However, this change did not necessarily reflect convergence in provincial emissions standards (Olewiler 2006, 133). A variety of factors, including federal regulations, affect ambient levels of air pollutants.

In examining air pollutants, Olewiler did not find evidence of competitive reductions in environmental regulations: Canadian environmental regulations change very infrequently. Given accumulation of scientific knowledge and technological advances, we should expect environmental regulations to be revised regularly. Most Canadian environmental laws are in place for decades before being revised. (A 2000 review of Ontario air standards noted that 'many were set more than twenty years ago' (Ontario 2000).)

Olewiler found limited evidence of occasional races to the top but added 'in a number of cases, there is a regression to national guidelines or standards, at times to a lower level than some provinces initially promised' (Olewiler 2006, 114). For example, in the late 1980s, several provincial governments made promises about

---

14   Ontario, Quebec, British Columbia, Alberta, Saskatchewan and Manitoba.

stringent dioxin regulations for pulp mills, which they subsequently backed down from. The federal standard, arrived at through federal–provincial cooperation, permitted significantly higher dioxin emissions than the standards which Alberta, BC and Quebec had initially announced.

The finding of infrequent revisions to Canadian environmental standards is consistent with the conclusion that provincial environmental standards are 'stuck at the bottom'. Olewiler noted that, even though the regulations have not been made less stringent, they may nonetheless be too lax relative to public preferences and the cost of environmental harm. Regulatory competition might be occurring, but not in the form of races to the bottom. The regulations are at a stable equilibrium but that equilibrium level is less than the optimal level.

Innovation? What Innovation?

If provincial environmental standards are indeed 'stuck at the bottom', this is contrary to a major claim in favour of decentralization: that it permits innovation and learning. It has been suggested decentralized governance may even stimulate technical progress. Ideally, individual states innovate, other states learn from that experience and as a result, innovations diffuse rapidly.

To assess these claims, political scientist Barry Rabe compared Canadian and American environmental policy regimes in the 1990s, measuring the degree of innovation in each system (Rabe 1998). He examined four indicators: 1) an emphasis on pollution prevention versus control; 2) the use of integrated permitting to reduce cross–media transfers; 3) the role of information disclosure and 4) outcome versus output measurement. He selected four provinces and four states to maximize regional diversity and variation on policy style.[15]

Rabe noted that provincial governments in Canada have all of the powers that American state governments want, as well as almost complete freedom from federal, if not judicial, oversight. For advocates of decentralization, these are the conditions which lead to experimentation and innovation and the kind of creative problem solving which makes for effective environmental policy. But Rabe concluded:

> The underlying premise behind most proposals to decentralize environmental policy is its presumed capacity to unleash creative, 'bottom–up' energies. But Canada's three decade experiment with widespread delegation of most environmental functions to its provinces reveals little subnational innovation. Instead, most provinces adhere to medium–based, pollution control–oriented regulatory systems constructed in the 1970s, appear eager to bend existing regulations to satisfy the overriding imperatives of economic development, provide minimal enforcement or monitoring of regulated parties and engage in

---

15 The cases were: Alberta, Manitoba, Newfoundland and Ontario; Arizona, Minnesota, New Jersey and Oklahoma.

minimum policy learning or idea diffusion with neighbouring provinces or the federal government.

The American states he studied were far more innovative and more likely to share ideas than Canadian provinces were. This suggests that the constraints imposed by minimum federal standards are more conducive to innovation than the complete autonomy enjoyed by Canadian provincial governments.

Economists are particularly keen on policy innovations which harness market forces for environmental protection, such as pollution taxes and emissions trading. Ironically, although the economist who started post–World War II interest in these instruments was Canadian (Dales 1968), economic instruments for environmental protection have been little used in Canada (Cassils 1991). Not only do Canadian jurisdictions not have pollution taxes or tradeable permits, they frequently do not charge consumers and businesses the full cost of water and sewer services. A few cities have effluent fees for discharging to sewers, but they are not high enough to create incentives for pollution reduction. Many Canadian cities do not charge consumers the full cost of water, a subsidy which encourages waste. In the mid–1990s, there were still 10 million households (a substantial portion of the Canadian public), including many in cities, without water meters (National Roundtable on the Environment and the Economy 1996). In the absence of metering, these households face no incentive to conserve water. The OECD's 2004 *Review* of Canada's environmental performance identified full cost pricing, particularly for water, and the use of economic instruments as priority areas needing improvement (OECD 2004b).

None of the data presented here supports the claim that a highly decentralized system will stimulate innovation. It is consistent with claims that decentralization leads to diversity of regulatory outcomes but the average performance is worse than the minimum in jurisdictions with floor standards. Furthermore, there is nothing to suggest that Canadian jurisdictions are leaders in environmental standards and protection, particularly in the international context. Minimum national standards are not fatal to innovation; the comparison of the Canadian and American cases suggest they help 'make the world safe for innovation' by limiting fears of competitive disadvantage.

Appropriate Adaptation to Local Environmental Conditions

Another argument made for decentralization is that it permits environmental protection tailored to local conditions. However, the available evidence on provincial regulations shows little propensity to tailor regulations to local environmental capacity or sensitivity. An assessment of effluents from 26 pulp and paper mills on Canada's east and west coasts found no adjustment was made for the assimilative characteristic of the receiving waters, such as depth and water flow. For example, researchers note that the Upper L'Etang estuary became 'incapable of supporting benthos and fish shortly after start up of the ... mill in 1970' (Colodey and Wells 1992, 206).

The finding here is similar to that for innovation. Decentralization is merely a permissive condition for tailoring to local environmental conditions; while it may permit tailoring, that does not mean it produces it. Just because lower level jurisdictions *could* set standards tailored to local conditions does not mean they actually *will*. The claim that national standards prevent appropriate local variation is based on a caricature: that national standards are uniform standards. However, where national standards for pollution control exist, they are overwhelmingly floor standards, not ceilings. For example, the federal regulations under the Canadian *Fisheries Act* did not prohibit provinces from setting more stringent standards, if they wished.

Representation of Public Preferences, Legitimacy and Accountability

Another claim in favour of decentralization is that lower levels of jurisdiction are better at representing preferences and that they are more accountable, hence more legitimate. There are some opinion polls showing respondents' opinion about environmental quality in their Canadian region or province. However, with regard to preferences, it is difficult to determine whether the environmental actions of provincial governments correspond to the preferences of their citizens. Similarly, it is difficult to determine if Canadians are relatively more satisfied with the environmental policies of the federal government or their provincial government. Available polls are insufficient to definitively resolve these questions. Polling results on preferences for federal or provincial authority are contradictory. Furthermore, Canadians may not be aware of which level of government has jurisdiction over what.

Poll results about local pollution   Polling results in Table 7.1 above shed some light on respondents' satisfaction with environmental quality in their province. Respondents clearly differentiated between the biggest environmental problems in their province versus the biggest national problem. Thus, in 1981 Western Canadians ranked acid rain much lower than Ontarians, who were affected by acid rain. The high proportion of Quebeckers (46 per cent) identifying water pollution as the major problem in the province strongly suggests Quebeckers were dissatisfied with water quality, but provides no information on whom they held responsible for this state of affairs.

Between 1970 and 1989, Canadians were polled about whether or not pollution was a problem in their area (Bozinoff and MacIntosh 1989). In 1970 and 1975, 56 and 53 per cent respectively answered yes. In 1980 and 1985, this proportion fell to 46 per cent but had soared to 67 per cent by 1989. As shown in Figure 7.1, in 1989, majorities in all regions (except the Prairies) answered yes. The proportions for the province of Quebec and Montreal were 81 per cent and 80 per cent respectively. Thus, while levels of concern fluctuated over time, concern was relatively greater in Quebec than other regions for both polls taken in the 1980s.

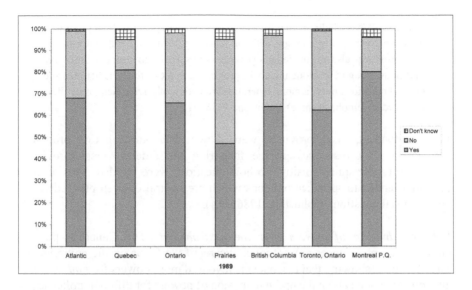

Figure 7.1 Is pollution still a problem in your area (Canada)?
Source: Bozinoff and MacIntosh 1989.

Poll results on the federal/provincial division of powers The limited public opinion data on the public's preferences for federal versus provincial action yield ambiguous findings. In a 1980 poll, 54 per cent of those polled about current efforts at constitutional reform favoured increasing the powers of the federal government AND also the powers of the provinces (Hastings and Hastings 1982, 90). (This contradictory result echoes a famous joke in Quebec: what a real Quebecker wants is 'an independent Quebec, in a strong Canada' (Deschamps 1977).[16]

The same 1980 poll asked respondents whether a new constitution should result in their own provincial government having (1) more power (2) less power or (3) an unchanged amount of power. The majority (50 per cent) favoured an unchanged amount of power for their own provincial government, with 36 per cent favouring more power and 10 per cent preferring less power. These results do not present a coherent viewpoint on jurisdictional preferences.

An overview of Canadian opinion polls from the mid–1970s to the early 1980s (a period of high constitutional conflict) produced these findings:

> [g]eneral questions on conflict and on the division of powers between the federal government and the provinces yield ambiguous results ... Majorities agree that provincial governments are more responsive than the federal government and, accordingly, that the provinces should have more power and that Ottawa should

---

16 '[...] le vrai Québécois sait qu'est-ce qu'y veut. Pis qu'est-ce qu'y veut, c't'un Québec indépendant, dans un Canada fort' (translation mine).

have less ... Overwhelming majorities in *each* province agree that their provincial
government should 'get tough' with the federal government. But majorities in
most provinces also see the federal government as 'their' government and agree
that Ottawa should 'get tough' with the provinces. Some of the jurisdictional and
conflict questions mask a more generalized rejection of government, regardless
of its order' (emphasis mine) (Johnstone 1986, 215).

Thus, respondents in all provinces were shown to favour their own provincial
government. But responses on the favouring the federal versus provincial
government were quite sensitive to how questions were worded. Less abstract
questions tended to 'produce opinions which are neutral or even pro–federal on
jurisdictional questions'(Johnstone 1986, 214).

Polls on division of powers over the environment   The Canadian public's
jurisdictional preferences are not much clearer if we focus on the environment.
The 1980 poll (showing that respondents favoured more powers for *both* levels of
government) asked about the optimal division of powers for different policy areas.
A majority of those polled (52 per cent) responded that environmental protection
should be controlled equally by the provinces and the federal government, while
the remainder were split equally between those favouring federal control (24 per
cent) and provincial control (23 per cent) (Hastings and Hastings 1982, 1982).

Results from a 1982 survey were somewhat more pro–federal. In a 1982
survey asking respondents which level of government should have the primary
responsibility of paying for environmental protection, 48 per cent chose the
federal government while 36 per cent chose provincial government (Johnston
1986, 200). Even those respondents with 'hardly any confidence' in the federal
government, favoured the federal government on this question (51 per cent over
34 per cent for the provincial government). In all regions, a plurality thought the
federal government should have primary responsibility for the environment, with
exception of Quebec where 47 per cent favoured provincial government versus 38
per cent favouring the federal government (Johnston 1986, 203).

A 1987 poll also showed pro–federal responses. Thirty–two per cent of
respondents indicated that the federal government had 'primary responsibility for
environmental protection', whereas only eight per cent said provincial government
did. ('Individuals' and 'industry' were also possible answers in this survey). It is
not clear if these figures reflect knowledge of actual jurisdictional responsibility
or whom respondents thought *should* be responsible (Runnals 1992). By 1990, the
proportion holding the federal government primarily responsible fell to 22 per cent
and those holding provincial government responsible fell to four per cent, which
still leaves a plurality holdings the federal government responsible.

The same 1987 and 1990 polls asked respondents to rate the environmental
performance of the federal and provincial governments. Between 1987 and
1990, the proportion who gave the federal government a 'poor' rating rose from
20 to 45 per cent. However, in the same time period, the proportion giving their

own provincial government a poor rating rose from 15 per cent to 36 per cent. This would be consistent with a general dissatisfaction with all governments. Unfortunately, these polls did not assess respondents' factual knowledge about what the responsibilities of each level of government were.

Seventy polls taken between 1985 and 2006 asked a sample of Canadians '[g]enerally speaking, do you approve or disapprove of the way the current federal government is handling the environment?' (Canadian Opinion Research Archive). The results showed that, between 1985 and 1993, majorities ranging from 50 per cent to 66.7 per cent disapproved, peaking in 1989. Between 1994 and 2004, slim majorities approved. However, this question reveals nothing about what actions those polled approved or disapproved of at that point in time. A 1990 poll asked respondents an open–ended question 'What should the federal government be doing?' The two most popular responses (at 19 per cent each) were 'pass stricter laws' and 'enforce laws/get tough' (Runnals 1992).

Although these poll results are ambiguous, they do not offer evidence of satisfaction with existing environmental quality in the respondent's own province. However, the polls reveal little about whom the respondents blame for this state of affairs. The findings on the public's preferences for which level of government should be responsible are contradictory. They do not show consistent support for greater provincial powers in the environmental policy. Nor do these poll results support a claim that Canadians consider provincial governments to be more responsive on environmental policy than the federal government.

The Record of Interprovincial Cooperation

In both the United States and Switzerland, frustration with decentralized environmental governance led to calls for horizontal cooperation, both formal and informal. Similarly, in Canada, the absence of a strong federal role in environmental policy has left a vacuum, a vacuum which interprovincial cooperation is supposed to fill. This section focuses on the Canadian experience with interprovincial cooperation, particularly through the primary institutional forum, The Canadian Council of Ministers of the Environment (CCME).

Formalization and Institutionalization of Cooperation

As noted in earlier chapters, both the American and Swiss constitutions permit subnational governments to negotiate treaty–like contracts with each other, to address problems that are transboundary in nature. No similar legal contracts have emerged in Canada. Furthermore, despite the great importance of interprovincial cooperation in the Canadian model of 'executive federalism', interprovincial agreements are not legally enforceable. Some legal scholars question whether the federal and provincial governments have the authority to enter into legally binding intergovernmental agreements (Lucas and Sharvit 2000, 133). In general, in comparison to other federations, the institutionalization of Canadian interprovincial cooperation remains low.

A comparative study of Intergovernmental Arrangements (IGAs) between subnational units within federations found that most Canadian IGAs (for example, on social policy) showed low levels of institutionalization. In particular, Canadian IGAs lacked the key features of the highly institutionalized IGAs found in some other federations: majority decision–making and the capacity/willingness to draft legally binding agreements (Bolleyer 2007, 231). In comparison to other IGAs in Canada, the Canadian IGA on the environment (the CCME) showed a medium level of institutionalization.

The CCME was characterized as having a medium level of institutionalization because, since the 1980s, it has had a large and well resourced secretariat. Given this high level of administrative capacity, the study's author found it striking that this secretariat's role is limited to facilitating information exchange (Bolleyer 2007, 232). Based on interviews with CCME officials, Bolleyer concluded that protection of provincial autonomy was a major priority in CCME activities, rather than facilitating interprovincial cooperation. In contrast to the Canadian provinces, the Swiss cantons actively pursued intercantonal cooperation with the explicit purpose of *discouraging* federal encroachment. Cantons were not concerned about the loss of individual autonomy that follows from such a cooperative approach, because they saw horizontal cooperation as strengthening the hand of cantons collectively, *vis–à–vis* the Swiss federal government (Bolleyer 2006b, 23). Perhaps this pragmatic stance has failed to emerge in Canadian IGAs because federal encroachment seems unlikely.

A dominant theme in Canadian environmental policymaking has been consultation and co–operation. While interprovincial cooperation on the environment has occurred since the 1960s, this cooperation has resisted legal formalization and higher levels of institutionalization. Although tremendous time and effort have been expended on interprovincial cooperation, this does not mean that cooperation has shown itself to be very effective. It is the only remaining option when individual provincial governments are unwilling to act and the federal government lacks the capacity, or chooses not to antagonize the provinces. How does the Canadian Council on Ministers of the Environment work? And how well has it worked?

The CCME and Joint Decision Trap   Because the Canadian federal government has not unilaterally set environmental standards since the 1970s, if the federal government sets standards at all, it adopts the standards reached by the CCME. The CCME makes decisions by consensus. This type of two level decision–making dynamic is what Fritz Scharpf termed the Joint Decision Trap. The trap was defined by two key conditions:

- that central government decisions are directly dependent upon the agreement of constituent governments; and

- that the agreement of constituent governments must be unanimous or nearly unanimous (Scharpf 1988, 254).

Furthermore, Scharpf identified several pathologies resulting from this type of decision–making, most importantly, lowest common denominator outcomes.

Scharpf claimed that, compared to decision–making in a unitary political system, the joint decision trap was suboptimal, but stable. The only way out of the trap was a transformation of the decision–making system. Others have claimed that the effects of the trap are mitigated if the central government has a credible option of exit; that is, if the central government can at least threaten to make a decision without being subject to the unit veto of subnational governments (Blom–Hansen 1999, 35). If the CCME represents a Joint Decision Trap, do we observe the pathologies that Scharpf identified?

The CCME represents all 14 environment ministers in Canada, one each from the federal, territorial and provincial governments. Established in the 1961, as the Canadian Council of Resource Ministers, the CCME had been largely dormant for the latter half of the 1970s. The 1980s saw the reanimation of the CCME: in 1988, it gained a larger budget and a permanent secretariat in Winnipeg. It devoted itself to trying to formulate national measures for a variety of environmental policies.

Effectiveness of the CCME   A 1995 study of the CCME's effectiveness found little evidence that the CCME has had an independent effect on improving environmental quality (Hillyard 1995). The CCME emphasizes maintaining harmonious relations among governments. The study found that the CCME voluntarily limited its ambit, ignoring some areas of environmental policy and producing strictly voluntary codes of practice and guidelines in all other areas (Hillyard 1995, iii). Despite the stated aim of developing national strategies, the CCME's guidelines do not result in consistent policies across Canada. The provinces do not necessarily implement the CCME recommendations that they themselves helped negotiate. Rather, provincial governments pick and chose which guidelines to adopt, if any.

In November 1993, the CCME's Council of Ministers announced that harmonization would be the CCME's top priority over the next few years (CCME 1994, 1). Harmonization appeared to mean different things to the provinces and the federal government, however. Federal bureaucrats wanted the Harmonization Initiative to create minimum national standards through inter–provincial cooperation. One of the stated objectives is:

- to develop and implement consistent environmental measures in all jurisdictions, including policies, standards, objectives, legislation and regulations (CCME 1998, 1).

However, the final outcome, the 1998 Canada–Wide Accord on Environmental Harmonization, fell short of this goal.

The provinces and industry argued that the rationale for the Accord should be the elimination of duplication and overlap (CCME 1998, 1). In 1978, the provincial premiers had identified overlapping federal and provincial environmental regulation as a problem and requested a study. The resulting study on regulation, covering several industries, found no evidence of overlap or duplication (Thompson 1980). A 1997 report to the House of Commons Standing Committee on Environment and Sustainable Development noted that no evidence of duplication or overlap was provided.[17] The president of Centre patronal de l'environnement du Québec (Quebec Business Council on the Environment) cited federal and provincial environmental impact assessment processes as one example of duplication and overlap (Cloghesy 1998).

The House of Commons Standing Committee heard testimony from political scientists and environmental groups who argued that the primary goal of the Accord was turf protection, not environmental protection. The main objective was keeping the federal government out of provincial areas of jurisdiction and limiting its jurisdiction. For example, both the second and fourth Objectives of the Accord emphasize having only one level of jurisdiction of government responsible for any particular role and eliminating overlap. Environmental groups and other NGOs were strongly opposed to the Accord; industry representatives were far more positive about it in their testimony before the Committee.

Characteristics of the accord on harmonization   The final Harmonization Accord resembles an international treaty or one of the old US interstate pollution compacts. Any government may withdraw from the Accord with six months notice. Its provisions are vague and unconstraining. The Accord contained Subagreements on setting Canada–wide Standards (CWS) and inspection (replaced in 2001 by a Subagreement on Inspections and Enforcement). The 2001 Subagreement focuses on clarifying lines of responsibility and the administration of delegated enforcement, rather than any clear commitment that the signatories will be held accountable to the public for enforcing their regulations. It provides for information sharing with other jurisdictions but does not mandate public transparency.

The province of Quebec never signed the Harmonization Accord and hence does not sign on to any Canada–wide Standards. Quebec has, however, indicated that it will behave in a manner 'consistent with the standards'(CCME 2000). Given that many of the standards involve action plans unique to each province, it is not clear what this means. Furthermore, any province may withdraw from a Canada–wide Standard upon three months notice. There is no indication of what, if any, penalty is incurred by withdrawing *without* three months notice.

---

17   (Canada. House of Commons. Standing Committee on Environment and Sustainable Development 1997, 14).

Setting Canada–wide Standards under the Harmonization Accord

The Harmonization Accord was accompanied by a work plan for negotiating Canada–wide Standards (CWS) in a variety of areas. Compared with the American or EU regulatory framework (which cover 100s or 1000s of substances), the number of pollutants addressed is very limited. Nonetheless, the process of negotiating Canada–wide standards has been slow. In 1998, the Harmonization Accord identified nine CWS to be negotiated. As of 2006, 13 Canada–wide standards had been concluded. Some are ambient standards; others are targets for emissions cuts. A few CWS are emissions standards for industrial sources.

This section will examine some of the Canada–wide standards that have been published to date. The resulting CWS are not particularly stringent. There is no requirement that these standards be adopted in provincial legislation, nor that they be enforced. Thus, although the standards ostensibly serve the function of 'national standards', they are not equivalent to the binding minimum standards found in US or EU regulations. The focus here will be on two ambient air CWS (ozone and particulate) and also the CWS for dioxin from waste incineration.

Canada–wide standards for ambient air quality   The scope of the Canada–wide Standards for ambient air quality is quite limited in international comparison. There are two CWS for ambient air quality: ozone and particulate matter (PM2.5). The US *Clean Air Act* set ambient air quality standards in 1970 and these been revised repeatedly since. As of 2008, the US has six ambient quality standards: carbon monoxide, lead, nitrogen dioxide, particulate matter (PM10) and (PM2.5) and ozone. The World Health Organization published six ambient standards in 1987, including sulphur dioxide, not found in the US standards. Between 1980 and 1992, the European Union set four ambient air quality standards: sulphur dioxide, nitrogen dioxide, lead and ozone.

While the Canada–wide standards for ozone and PM2.5 are approximately as stringent as the US standards, the policy consequences of the Canadian and American standards are very different. The Canada–wide Standard for ground level ozone, set in 2008, is 65 ppb (averaged over 8 hours), which is in line with the recommendations made to the US EPA by a scientific advisory panel, in 2007. The CWS is more stringent than the new 75 ppb standard the EPA announced in 2008, which replaced a standard of 84 ppb (Wald 2008). The more stringent standards were recommended because higher levels of ground level ozone are associated with increased death rates among the elderly and those with lung diseases. Based on a study of eight Canadian cities, Health Canada estimated that air pollution in those cities contributes to 5,900 premature deaths annually (Environment Canada et al. 2007, 2).

However, while the numerical values of the US and Canadian ambient standards are similar, their policy consequences are very different. Ambient standards, by themselves, are unenforceable; they only have an effect in conjunction with other measures (such as regulations or taxes) that *limit the emissions* causing ambient standards to be exceeded. The OECD 2004 *Review* noted that, under the Canada–wide Standards, there are no penalties for noncompliance nor any plans for distinguishing between areas that meet ambient standards and those that do not yet meet them (OECD 2004a, 34). In contrast, the US distinguishes between 'attainment' areas, generally in compliance with ambient standards, and 'nonattainment' areas, which exceed ambient limits. The US EPA has policies tailored to air quality 'nonattainment' counties, to improve air quality, as well as policies to prevent 'attainment' counties from slipping into 'nonattainment' status. In particular, non–attainment counties face federal restrictions on new highways and industries (Wald 2008). Thus tightening the ambient standard increased the number of non–attainment counties to 345. The EPA estimated that bringing these counties into attainment would cost US $8.8 billion (Wald 2008).

CWS on dioxin from incineration of municipal waste   In the 1990s, the burning of waste was identified as a major source of dioxin emissions to the air. Under the *Canadian Environmental Protection Act*, dioxins are designated for 'virtual elimination' from the environment, because they are carcinogenic and highly toxic. Although Canada has made progress in reducing dioxins, these efforts have been hamstrung by the CCME's unanimity requirement. In 2000, the CCME announced plans for a Canada–wide Standard for dioxin, which the provinces would be responsible for implementing. That year, the CCME announced a numeric emissions limit for municipal solid waste (MSW) incinerators, 80 pg I-TEQ/m^3 (or 0.08 ng) (CCME 2000b, 7). Existing municipal solid waste incinerators were to meet the standard by 2006.

However, one of Canada's largest remaining sources of dioxin emissions to the air was *explicitly excluded* from this standard (CCME 2000c). Investments were made to achieve marginal improvements in dioxin emissions from modern MSW incinerators. Meanwhile, in the province of Newfoundland and Labrador, as of 1999, 45 teepee or conical waste combusters continued to operate. These very low technology, low temperature combustion plants are not much more advanced than open burning of garbage and result in very high levels of dioxin emissions for each unit of garbage burned (CCME 1999).[18] (They also contribute to mercury emissions.)The justification for excluding the teepee burners was that it was impossible for these burners to meet the Canada–wide standard.

Although Newfoundland and Labrador is not very populous (about 500,000), its teepee burners produced *more* dioxin than the nine large MSW incinerators in the rest of Canada put together (75 g TEQ/year versus 67 g TEQ/year) (UNEP 1999, 66). This discrepancy became even greater because 62 of those 67 g TEQ/year came from a single incinerator in Quebec. Once this source was modernized,

---

18   Whereas modern MSW incinerators operate at temperatures of at least 800°C (1470°F), teepee burners operate at far lower temperatures. In 2006, a man was rescued after spending about five minutes in an operating teepee burner that he had fallen into: 'I wasn't going to burn to death. I was going to cook' (Brautigam 2008).

emissions from *all* Canadian MSW incinerators fell to 7 g TEQ/year in 1999. From then on, emissions from the teepee burners were *ten times* those of the all the MSW incinerators in the rest of Canada. As of 1999, Newfoundland teepee burners were thought to account for 42 per cent of total Canadian dioxin emissions to the air, even though they burn a tiny fraction of the waste of the modern MSW incinerators (CCME 1999, 2).

In a 2000 statement addressing the Canada–wide Standard, the government of Newfoundland and Labrador pledged to 'review the use of conical waste combusters'(CCME 2000c). In the 2004 final agreement, the conical waste combusters were to be phased out by 31 December 2008. In April 2008, a majority of those 45 conical waste incinerators (25) were still operating.[19] In October 2008, the Municipal Affairs Minister acknowledged that only three or four of the remaining 25 incinerators would be closed by 2009 (Brautigam 2008). Extensions would be granted 'on a case-by-case basis' for the still operating incinerators.[20]

What effect has the Canada–wide Standard setting process had? In international comparison, Canada has been slow to regulate dioxin emissions from MSW (Weibust 2005, 52–3). From the dioxin example, it is difficult to conclude that the Canada–wide Standard caused the government of Newfoundland and Labrador to act on dioxin emissions any sooner than it would have *absent* the Canada–wide Standard. Eight years after the CCME dioxin standard had been negotiated (2000), this source of dioxin continued to pollute, unabated and unregulated. The outcome of the CWS process was that the dioxin source which was most polluting (by weight of garbage), and eventually, by total dioxin emissions, was regulated last.

The Joint Decision Trap and the CCME

How does well does the Joint Decision Trap typology fit the CCME? The inability of the central government to make a decision without virtual unanimity from constituent governments generally describes the how the CCME makes decisions and how the federal government only adopts standards that have been negotiated by the CCME. There are a few points of divergence from Scharpf's model. First, Scharpf described policy areas where joint decisions were binding. The CCME's standards are clearly not binding, although they would technically be legally binding, if embodied in a federal regulation. Second, under Scharpf's typology, once decisions in a particular domain had to be made jointly, unilateral action or defection by subnational governments was no longer possible. Because cooperation through the CCME does not limit provincial autonomy, individual provinces could continue to set their own standards (or not set any standards at all).

Do we observe the pathologies that Scharpf attributes to the Joint Decision Trap? He points to suboptimal outcomes as one result. Certainly, many of the CCME's standards are relatively weak compared to standards set in other jurisdictions. One

---

19   'NLEIA approves of planned incinerator shutdown' 2008, A14.

20   'Garbage Incinerators to Get Extension, Minister Says' 2008.

of the causes of suboptimal decisions is that bargaining under unanimity is unable to address distributional issues and fairness (Scharpf 1988, 265). This seems to characterize the outcome in the dioxin case and other Canadian environmental standards, such as sewage treatment.

Newfoundland and Labrador was clearly concerned about the cost of closing down a large number of its highly polluting tepee burners, which would be costly for the small municipalities that operated them. However, under the CCME framework, the only accommodation that could be granted was exemptions and delays. The agreement on a uniform national standard and a timetable for shut down could have been accelerated, had the CCME decision–making process allowed for financial compensation to enable the province to comply more promptly by subsidizing alternatives for the affected municipalities. As a national policy, it would have been more efficient to allocate resources to shut down teepee burners, rather than to invest in marginal improvements to larger incinerators that were already relatively clean.

However, the fragmentation of the CCME process essentially rules out this kind of national calculus for efficient measures. Furthermore, the CCME process does not give rise to compensation or side–payments to facilitate compliance with standards. In theory, such side agreements should be easy to facilitate in a federal system because the governments interact on such a wide range of issues (Axelrod and Keohane 1986). It should be possible to create tradeoffs in other fora that compensate for the costs in this one. However, it is possible that in Canada, the tremendous number and specialization of such fora actually prevent tradeoffs. Each federal–provincial forum is essentially a silo and operates in its own highly specialized policy area. Any brokering of side–payments across provinces would almost certainly have to take place at the level of the premiers.

Conclusion

It is very difficult to reconcile the evidence presented above with the image of Canada as an environmental leader. Comparative data on environmental quality show that Canada pollutes relatively more than other jurisdictions, both by population and by unit of economic output. Compared to other jurisdictions, Canada is slow to act in addressing environmental problems nationally. Because there are no national floor standards to bring up the rear, many of the worst environmental problems, such as untreated sewage, persist decades after they have been addressed in the US or the European Union.

The Canadian pattern of environmental standard setting is consistent with what was observed in the US and Switzerland, prior to a stronger federal role in those federations. Environmental regulations were relatively few and varied substantially across subnational jurisdictions. It was difficult to obtain data on local environmental quality, local regulation or the enforcement of those regulations. In Canada, there is no evidence of a race to the bottom; however, the outcome is supoptimal for environmental protection. The few standards that exist are relatively lax and change rarely. Rather than racing to the bottom, standards remain stuck at the bottom.

With a few exceptions, Canadian federal environmental policy has been characterized by a deference to provincial governments which borders on abdication of responsibility. Not only has the federal government deferred to provincial legislation in almost every instance, it has ceased to enforce its own legislation in many cases as well, leaving it to the provinces to do so.

In trying to overcome the limitations of a very decentralized regulatory system, successive federal governments have tried to secure the benefits of minimum national standards. On a few occasions, the federal government has made brief forays into setting minimum national standards. However, since the 1980s the federal strategy has been to reach national standards by means of interprovincial cooperation. This reliance upon interprovincial cooperation falls prey to the Joint Decision Trap, which gives individual provinces a veto in setting national standards. The consensus based process results in fewer and less stringent standards than found in more centralized jurisdictions.

The extensive provincial role in setting, implementing and enforcing national standards introduces four points of slippage in the national regulatory system. These points of slippage exert significant downward pressure on Canadian environmental protection, helping to keep Canadian environmental policy 'stuck at the bottom'. First, consensus decision–making tends to result in less stringent standards, because it contributes to lowest common denominator decision–making (Scharpf 1988). Secondly, because provincial standards are infrequently updated and cooperative standard setting is slow, the passage of time ensures that Canadian standards become outdated compared to those of other jurisdictions that are able to make decisions more quickly.

Third, despite the very significant role that each province plays in determining national standards/guidelines, there is very substantial variation in implementation of the guidelines. Extensive provincial participation in standard setting does not guarantee buy–in by the participants: Canada–wide standards negotiated by the Canadian Council of Ministers of the Environment are not binding on the parties. Nor are provincial governments obliged to put the Canadian Drinking Water Quality Guidelines into legislation. The final point of slippage is enforcement. Provincial governments are not compelled to enforce their own standards, assuming they actually have binding standards rather than mere guidelines. In theory, when provincial governments have agreements to enforce federal standards, such as the *Fisheries Act* regulations, they have an obligation to do so. However, because the federal government has chosen not to audit provincial enforcement of federal regulations, it is very difficult to determine the extent to which those regulations are being enforced.

From the evidence presented above, there is little to suggest that noncentralized environmental policy fulfills the promise that its advocates claim. In the Canadian context, a weak federal role has not lead to innovative policy, characterized by nuanced responses tailored to local environmental conditions. Nor is there evidence to suggest poor environmental quality and government inaction represent the will of the electorate. It is quite possible that the electorate has little idea of what actually goes on. In the Canadian case, noncentralization is accompanied by serious gaps in information, particularly at the subnational level: a lack of data on environmental quality, policy performance, standards and their enforcement. The absence of performance data, combined with the extreme fragmentation of the system makes it difficult for specialists to get 'the big picture', never mind the average citizen.

The federal government has launched numerous initiatives aimed at reaching consistent national policies, by means of intergovernmental cooperation. None of these attempts has resulted in binding standards, nor have they resulted in standards which will be uniform nation wide. Provinces have asserted their prerogative to set their own standards, even after participating in prolonged exercises on cooperative standard setting. There is also nothing to prevent any province from withdrawing from a negotiated standard.

The emphasis on harmony and flexibility in intergovernmental negotiation is equally unproductive. It produces accords which do not demonstrably change any government's incentives and are not constraining on governments in any way. It suggests that cooperative, unanimous decision–making mechanisms are not sufficient to allay the fears of interprovincial economic competition. Although provinces have the opportunity to negotiate standards which are binding, they chose not to do this. Even if everyone were bound by the agreement, the participants would prefer not to sacrifice their autonomy.

While these findings are significant for cooperation in federal systems, they also have broader implications for international cooperation. The Liberal institutionalist view is dominant in international environmental politics. It predicts that in the presence of favourable contracting conditions, high levels of concern and capacity, environmental cooperation can be successful, without encroaching on sovereignty (Haas et al. 1993; Keohane and Levy). Canadian provincial governments have shown themselves to be unwilling to be bound, even by agreements that they themselves negotiate. The dense networks of interaction between the same governments on myriad issues seem to have little impact.

International lawyers go further, arguing that binding commitments and enforcement play little part in determining the effectiveness of agreements.[21] They argue that failures in compliance are often due to factors beyond the control of signatories and that they do not reflect opportunism and intent to renege on commitments. Canadian environmental regulations, particularly those passed by the federal government, have had periods of being enforced and periods of being unenforced. Evidence from the pulp and paper industry indicates that some provincial governments will avoid enforcing federal environmental law, when they have been delegated the authority to do so. Evidence on the behavior of firms

---

21   See for example (Chayes and Chayes 1995).

in this sector suggests that firms are more likely to comply with regulations which they know are being enforced. Firms and governments will cheat and are more likely to cheat if they think they can get away with it.

Chapter 8

The European Union: Setting Stringent Standards despite the Obstacles

The environment was not mentioned in the Treaty of Rome which created the precursor to the European Union. Yet, over time, the environment has come to be one of the European Union's core competences. The European Union is clearly not a federal state, yet many of the issues surrounding environmental standard setting in a federation also pertain here. The EU is not an ideal case for testing the efficacy of decentralized standard setting because its member states have never had uniformly high levels of concern about the environment, even prior to the accession of the Eastern European member states. The heterogeneity in preferences has only grown with successive waves of accession to the Union.

This chapter applies the findings from the previous three cases, to see if the same patterns obtain in the European Union. First, has a shift to standard setting at the European Union level led to more high standards, as the US and Swiss cases predict? If yes, the outcome cannot readily be attributed to EU wide popular pressure for stronger environmental protection. The EU's heterogeneity in preferences over environmental concern, however, does make it possible to isolate the effects of institutions and rules on environmental policymaking.

Second, what features of EU decision–making appear to make it effective? The US and Swiss cases showed that federal governments were able to set stringent standards, where uncoordinated subnational action as well as cooperation had failed. The Canadian case indicates that informal cooperation on standard setting has not been effective. In terms of governance, the EU occupies an intermediate position between centralized US and Swiss federal governments and the cooperative governance within federations. If governance within the EU has been effective, what differences between subnational cooperation elsewhere and EU decision–making can account for the difference? In particular, did consensus decision–making by the Council of Ministers result in lowest common denominator environmental policies at the European Union level? If not, then why has stringent standard setting been possible?

The chapter has seven sections. The first section presents data on preferences for environmental protection across the member states. The second describes environmental governance in the EU, including the application of the subsidiarity principle. The third describes the rules governing decision–making in the Council of Ministers. The fourth presents the EU's early environmental directives and assesses their stringency. The fifth presents findings from a study of regulatory convergence inside and outside of the European Union. The sixth section compares the findings from the EU from those of the classic federations presented above.

Preferences for Environmental Protection Across Member States

Historically, European Union member states have been quite heterogeneous in their preferences over environmental protection. Informal and formal analyses have classified member states as leaders or laggards. There is some dispute over which member states belong in which camp (Börzel 2003). An analysis of positions of member states in the Council of Ministers between 1980 and 1995, finds that Denmark, the Netherlands and Germany were clearly leaders, whereas Spain stood out for its consistent opposition to more stringent environmental standards, often on economic grounds (Weale et al. 2000, 96). These differences are also reflected in public opinion data. Eurobarometer data finds great disparities, which have grown over time. In both 1974 and 1993, there were substantial differences across member states in the proportion of respondents identifying 'the environment' as the most important problem facing member states (Weale et al. 2000, 96).

Environmental governance in the European Union

What kind of governance system is the European Union? It has elements of a federal system. To be considered truly federal, a system of governance must permit each level of government to be autonomous in at least some area of policy. If decisions at the lower level can be overruled by a higher level of government, then the system is merely decentralized and not truly federal.

In the European Union, this division of competencies is, in theory, subject to the subsidiarity principle: decisions should be made at the lowest possible level of government. The 1987 Treaty first set out the principle, in Article 130r(4), '[t]he Community shall take action relating to the environment to the extent to which the objectives referred to in paragraph 1 can be better attained at the Community level than at the level of the individual member states'. This was replaced by Article 3b of the 1993 Maastricht Treaty (now Article 5 of the current treaty), which states,

> [i]n areas which do not fall within its exclusive competence, the Community
> shall take action, in accordance with the principles of subsidiarity only if and in
> so far as the objectives of the proposed action cannot be sufficiently achieved
> by the member states and can therefore, by reason of the scale or effect of the
> proposed action, be better achieved by the Community.

This provision applies to all competencies, not just the environment. It should allow for maximum diversity of policy choices and keep policymaking close to the people. Because the environment is an area of shared competence (not exclusive EU competence), the subsidiarity principle is supposed to apply here.

The Amsterdam Treaty (1997) added an Annex, 'A Protocol on the application of the principles of subsidiarity and proportionality', containing 13 detailed points. The leading expert on EU environmental law Ludwig Krämer argued that this Protocol was added, not to develop EU jurisprudence but rather 'with the aim of containing as far as possible the threat of European integration which was perceived as reducing the amount of power that regional or national administration had up until then' (2007, 17).

In practice, the European Union's application of the subsidiarity principle bears only a family resemblance to the economists' criteria for determining the 'lowest possible level'. In the opinion of one observer,

> [i]t is tempting then to conclude that subsidiarity in the European context offers a relatively hollow promise to European local governments, even though theoretically the principle favours the local, as well as regional and national levels (Flynn 2000, 77).

The Commission's interpretation of subsidiarity appears to reject the subnational level as the appropriate locus for European decision–making, with the exception of federations within the European Union (Flynn 2000, 76–7).

Eleven years after the introduction of the subsidiarity principle, Ludwig Krämer wrote,

> I do not know of one single environmental measure since 1987 where the Council [of Ministers] has decided or even discussed whether a measure could be better adopted a Community level than at the level of Member States ... it was at least never expressly stated that reasons of subsidiarity played any role in the Commission's or the Council's attitude towards proposals (1998, 73).

He considered the determination of the appropriate level of jurisdiction to be a political question, not a legal or technical one (2007, 20). For example, he notes that 'better' is not defined in Art. 3b (now Art. 5); '... it could mean quicker, more effective, cheaper, more efficient, closer to the citizen (i.e. not too centralized), more democratic, more uniform, more consistent with measures in other parts of the industrialized world or the global or the European Community without these concepts being more precise' (2007, 19).

In the European Union, there is a strong preference for centralizing those policies deemed necessary for the operation of the Single Market. The European Union has defined these necessary policies much more broadly than economists would, however. Hence, a wide range of European Commission competencies are justified by the need to 'equalize the conditions of competition'. In the absence of barriers to internal trade or harmful environmental spillovers, economists consider this approach to be inefficient and excessively centralizing.

Because 'equaliz[ing] the conditions of competition' has come to be defined very broadly, the Commission has had a strong role in environmental policies where few environmental spillovers exist and where there is no harm in competition. Ludwig Krämer noted '... it may well be argued that the quality of the (local) bathing waters, drinking water or urban waste water are questions of a purely local nature and should, therefore, not be regulated at the EC level' (2007, 3). A prime example is the 1975 Directive 76/160 governing the quality of bathing water at beaches. The preamble to the directive refers to 'the harmonious development of economic activity throughout the community' and measures 'directly affect[ing] the functioning of the common market'.[1] Most economists would vigorously dispute this claim.

A cranky British opponent of the Bathing Water Directive pointed out the weakness of the competition argument:

> ... I have not heard a single person asking what right Brussels had in the first place to tell us how clean our seawater should be ... nowhere in [the Treaty of Rome] can I find any reference to sewage, snorkelling our sunbathing. Can the presence of bacteria in Blackpool be considered a threat to what Part Two of the treaty calls a 'harmonious development of economic activities?' Only, I suppose if you argue that it interferes with competition by giving an unfair advantage to rival tourist resorts in, say, Denmark or Portugal. But if the Danish or Portuguese resorts have that advantage, surely it is a fair advantage? You might as well issue a directive ordering restaurants of England to improve their cooking, on the grounds that this prevents their giving an unfair advantage to the restaurants of France.[2]

Because polluted beaches in the UK do not cause pollution in other countries, most economists would argue that the hygiene of beaches should be a wholly domestic, if not local, matter. In theory, it should be possible for the British to autonomously determine their preferred level of bacteria at the beach.

In a similar vein, the Preamble to the 1980 directive on Drinking Water Quality states that:

> ... whereas the disparity between provisions already applicable or in the process of being drawn up in the various member states relating to the quality of water for human consumption may create differences in the conditions of competition and, as a result, directly affect the operation of the common market; whereas

---

1   Council directive 76/160/EEC of 8 December 1975, concerning the quality of bathing water.
2   'The Spectator's View' 1990, 19.

laws in this sphere should therefore be approximated as provided for an article 100 of the treaty ...[3]

Again, this directive cannot be justified on the basis of controlling externalities. It concerns the quality of water as it comes out of the tap in European cities. This is a purely local public good, not a good that is traded across member states. Economists would dispute the claim that differences between cities on this parameter conferred any competitive advantage, much less an unfair one.

Despite these violations of the principle of subsidiarity, the environment remains, in theory, an area of 'shared competence,' with jurisdiction shared by the EU and member states. Member states have discretion in how they implement the legislation passed by the EU. The EU does not preclude member states from making autonomous environmental regulations, provided these do not conflict with existing EU legislation or impede the functioning of the Common Market.

Where the EU has promulgated a regulation, usually in the form of a Directive, member states are required to transpose this Directive into their own national legislation, in order to give it effect. Failure to properly transpose a Directive may be reported to the European Commission and may result in action by the European Court of Justice. The Court can apply penalties for noncompliance. If the Directive is deemed to have been properly transposed, the member states have discretion in how it is implemented. Since the *Francovich* decision by the European Court of Justice in 1991, European citizens have standing to sue 'if their rights are breached by infringements of Community law attributable to a Member State'(Jans 1996, 83).

Even though member states retain discretion in implementation, the EU's enforcement mechanisms appear to have a significant impact. A comparative study of compliance with similar rules made by the World Trade Organization, the European Union and Germany found that compliance with European Union directives was at least as good or better than compliance with either the WTO rule or German federal government policy. The finding that compliance with the EU directive was no worse than the national German measure was particularly surprising. The authors attribute the higher level of compliance to EU procedures for monitoring and enforcement (Zürn and Neyer 2005, 215).

Decision–making by the Council of Ministers

How have these Directives been passed? The EU's decision–making machinery has evolved over time, as the European Parliament has been given a greater role and the use of Qualified Majority Voting in the Council of Ministers has been extended. The Council of Ministers is comprised of the national ministers, one

---

3 Council Directive 80/778/EEC of 15 July 1980 relating to the quality of water intended for human consumption.

from each member state. The EU's decision–making procedure varies depending on the appropriate constitutional basis of the legislation in question, a determination which is not straightforward (Koppen 1993, 143).

The most significant feature, however, is the decision–making rule in the Council of Ministers, which still has the final say. Prior to 1987, decisions on the environment were made unanimously by the Council. In 1987, an environment directive (on automotive emissions) was passed by Qualified Majority Voting (QMV) for the first time (Arp 2005, 269). Under QMV, the value of the vote of each member state is calculated by a formula which partially reflects population. The *1987 Single European Act* (SEA) introduced QMV for those environmental regulations concerning the completion of the European common market (based on Art. 100A). After the 1992 Maastricht treaty, almost all aspects of environmental policy have been subject to QMV. Scholars have not identified significant impacts on the level of environmental standards following the 1987 extension of QMV (Knill and Liefferink 2007, 18–19). Despite the legal change allowing QMV, the Council of Ministers seeks to avoid formal votes and unanimous decisions are still preferred (Knill and Liefferink 2007, 88).

Legislative Program of European Union

Since the early 1970s, the scope of European Union environmental policy has expanded dramatically. Assessing these developments, two European scholars concluded that

> the output of this first phase of European environmental policy can by all means be regarded as a success. Within about a decade, a very substantial set of European environmental laws emerged with many important areas of environmental policy being regulated at the European level (Knill and Liefferink 2007, 9).

As of 1999, the number of European Union environmental directives was estimated at between 300 and 500 directives (Haverland 2003, 205). Not all directives are equally important; many are amendments of prior legislation.

In general, the EU was less able to craft regulations governing industrial sources than publicly owned pollution sources. Two framework directives intended to address industrial emissions from air and water respectively witnesses relatively few follow up 'daughter' directives setting specific standards. For example, the UK successfully fought a daughter directive to set limits for chromium emissions to water, as well as a proposal to regulate emissions from pulp mills (Golub 1996, 719).

Measures directed at sources of pollution likely to be publicly owned, such as sewage treatment plants or waste incinerators, have been stringent. The authors of *The Effectiveness of European Union Environmental Policy* claim that water quality moved to the forefront of the EU environmental policy agenda because the water sector, both supply and treatment, remain predominantly publicly owned in the EU. They argue that the water sector forms a much weaker lobby at the EU level than private corporations do (Grant et al. 2000, 158).

Since the early 1990s, the EU has largely abandoned setting emissions limits for industrial sources. The EU has replaced legislated emissions limits with the Integrated Pollution Prevention and Control Directive 96/61, which 'substantially increases the discretionary power of the member states with regard to the authorization of discharges of hazardous substances' (Pallemaerts 2005, 210). This has been regarded as a triumph for the traditional British approach to environmental regulation, which eschews numerical standards and emphasizes the context of local environmental conditions. This development was lamented by the head of the European Environment Bureau, an umbrella organization of European environmental groups, but praised by British industry (Golub 1996, 720). The shift from emissions limits to the Integrated Pollution Prevention and Control regime now makes it very difficult to compare standards across countries and across industrial sectors.

Stringency of EU Directives

Has the EU set stringent standards or have they simply reflected the lowest common denominator among member states? Scholars of European Union *product* standards agree these are clearly not lowest common denominator standards (Holzinger 1994). There is a consensus that harmonized product standards, such as those governing car emissions, have been set at high level (Vogel 1995). Thus the analysis presented here will focus solely on process standards to reduce air and water pollution. The analysis focuses on those directives passed prior to 1987, when the unanimity rule in the Council of Ministers was in effect. Under conditions of unanimity, proposed directives faced a higher hurdle than they do under QMV. A unanimity rule is generally considered less conducive to stringent standard setting than some form of majority voting.

The measure of regulatory stringency used here is the degree to which the directive requires a departure from the status quo. Directives which reflected the lowest common denominator among member states would require very few changes and little expenditure for implementation. If, on the other hand, implementation requires extensive expenditures, even for the older, wealthier member states, the Directive can be judged to set a higher standard, relative to existing national measures. Alternatively, if a directive explicitly requires reductions in total national emissions, compared to present emissions, it represents a change from the status quo.

Emissions to air In 1984, the EU passed the framework directive on the Combating of Air Pollution from Industrial Plants 84/360. This directive gave rise to the following daughter directives: the 1988 directive on Large Combustion

Plants 88/609 and directives on New and Existing Municipal Waste Incinerators 89/369 and 89/429.

The Large Combustion Plants directive was passed in order to reduce gases that were causing acid rain. The directive set emission limits for new combustion plants. Member states also pledged to cut total national emissions of SO_2, NO_x and suspended particulate from these plants by certain dates. Although poorer, newer member states were allowed to increase total national emissions of SO_2 and NO_x, most member states were required to make cuts. Belgium, Denmark, the United Kingdom, Germany, France, Italy and the Netherlands were all required to make cuts in SO_2 emissions in excess of 60 per cent over 1980 levels, by the year 2003. Reductions in total national emissions of NO_x had to be cut by between 20 to 40 per cent by 1998 in these same countries. The Large Combustion Plant directive required substantial national level cuts which represent a significant departure from the business as usual scenario.

In 1989, the EU passed directives on air pollution from New and Existing Municipal Waste Incinerators (directives 89/369 and 89/429). The first directives specified combustion conditions intended to reduce the production of dioxins/furans. Some member states with municipal solid waste (MSW) incinerators had no standards for emissions of dioxins/furans. Spain had no regulation, nor did Belgium which had 28 MSW incinerators at the time the first EU incineration directive was passed. In 1998, the EU set a target of cutting dioxin emissions from MSW incineration by 99 per cent from 1994 levels, by the year 2005. A 99 per cent reduction in anything must be seen as a stringent measure. The 2000 revision of the EU directive introduced emissions standards as strict as any in the world (Weibust 2005).

Emissions to water   The 1976 framework directive 76/464 was intended to reduce emissions of hazardous substances into water. Based on a German approach, it listed two types of substances. The most hazardous chemicals were on the 'Black list' and emissions of these substances were ultimately slated for 'elimination', In the mean time, the directive stated that authorization permits with specific emissions limits were to be issued to facilities emitting these substances. Only 17 of the 129 substances on the Black List were ever made the object of 'daughter' directives, with specific emissions targets (Pallemaerts 2005, 210). In a study of deregulation in the European Union, European industry sources consistently identified this directive as one of the most demanding they have to work with (Grant 1998, 147).

The 1980 Directive on Drinking Water 80/778 imposed substantial compliance costs on all member states. Even Germany, whose monitoring system was technologically advanced, had to acquire new measurement technologies to comply with the monitoring requirements of the Drinking Water Directive (Börzel 2003, 2). Along with the Urban Wastewater Treatment directive, this is one of the directives that has had the biggest impact on the water bills of British consumers (Byatt 1996, 665).

In 1975, the Directive on Bathing Waters 76/160 was unanimously adopted. It was intended to improve water quality at beaches and provide citizens with information about water quality. The directive set out sampling requirements and minimum standards for bacteria counts. It had the potential to be very costly for member states because the major determinant of bathing water quality is sewage outfalls to the ocean. Improving bathing water quality would require building sewage outfall pipes further into the ocean or improving the quality of sewage treatment. Member states were required to identify beaches used for swimming. In France, a similar national directive on water quality at ocean beaches had been on the books since 1964 (Wurzel 2002, 187).

The UK was generally critical of the directive. In 1975, the UK Department of the Environment predicted that the cost of complying would be £100 million (Jordan 2002, 118). The UK sought to minimize the impact of the directive by only listing 27 beaches (against thousands listed for Denmark, France and Italy), even though the UK had the longest coastline in the EU at that time (Moore 1992, 2). Persistent noncompliance with the Bathing Waters directive eventually led the European Commission to propose the Urban Waste Water Treatment Directive, to address the underlying problem of inadequate sewage treatment (Jordan 2002, 123).

One of the directives with the most far reaching implications has been the 1991 Urban Waste Water Treatment Directive 91/271. This directive requires that cities subject all their sewage to secondary, also known as biological, treatment. In some cases, it also requires improvements to sewer networks. Sewage systems discharging into areas judged environmentally 'sensitive' must employ more extensive tertiary treatment. Eventually, the directive will apply to all cities and towns with populations greater than 10,000 persons. These requirements represent major infrastructure investments.

The impact of this directive is best reflected in the costs of complying with it, which often require the construction or renovation of sewage treatment plants. The European Commission has estimated that, for the Central and Eastern European accession candidates, the cost of complying with this single directive would be between €80–110 billion, or €1057 per capita.[4]

Implementing the directive has imposed significant costs on older member states as well. Germany faced substantial investments in the former East Germany, estimated in 1993 to range between DM 35–40 billion. Even for the former West Germany, which had high standards for sewage treatment, implementation was estimated at €32.4 billion at €530 per capita for the period 1993–2005 (Kemp 2002, 86–88). The directive is considered to be one of the most expensive directives the United Kingdom has ever had to implement. In 1990, when the Directive was still a draft, the UK government estimated its total cost at £1.5 billion. This estimate

---

4   This estimate covers Bulgaria, the Czech Republic, Estonia, Hungary, Latvia, Lithuania, Poland, Romania, the Slovak Republic and Slovenia. 'Enlargement and the Environment: Questions and Answers'. DG Environment. May 2002.

proved to be far too low. By 1998, the government had revised its cost estimate for 1993–2005 to £8.9 billion (Kemp 2002, 243).

As of 2005, France and Spain had not yet fully implemented the directive. A 2005 report on the directive notes that the cost of implementation is the main reason for delays (EEA 2005, 5). Spain received significant subsides from the EU in order to support implementation, covering up to half of total Spanish investment in sewage treatment. Between 1993 and 2002, more than €3.8 billion were provided from the EU's Cohesion Fund, covering up to 85 per cent of individual plant investments (EEA 2005, 11).

Policy Convergence, Inside and Outside of the European Union

There is little to suggest that the above directives were set at the lowest common denominator. Many required very substantial changes on the part of the wealthier member states, and occasionally even 'leaders' such as Germany. Are there competing explanations for EU standards being set at a higher level than member state standards? One possibility is that member states are caught up in a global trend towards higher standards. This would imply that, over time, their standards would have risen with or without EU membership.

The findings of the ENVIPOLCON research project on Environmental Policy Convergence in Europe can shed light on this issue (Albrecht et al. 2005, Holzinger et al. 2008). One of the core research questions was: to what extent and in which direction can we observe a convergence of national environmental policies in Europe over the last 30 years? In one study, the researchers examined 40 environmental policies in 24 countries, over the period 1970 to 2000, including 21 policies which had quantitative standards. The sample contained 21 European countries, of which 14 were EU members, as of 2000. Among the policies covered were the following quantitative emissions standards:

- SO_2 emissions from factories
- dust emissions from large combustion plants
- industrial zinc emissions to water.

The presence or absence of a policy on industrial discharges to water was also part of the sample (Albrecht et al. 2005, 177).

The study employed various measures of policy convergence. Using a gap approach to measuring convergence, the authors were able to assess the direction as well as the degree of convergence. This approach to convergence measurement is necessary in order to detect races to the bottom, for example. On the basis of the gap analysis, the authors concluded '...the general picture is clear ... the average policy gap is decreasing over time for (nearly) all items, indicating an upward trend in policy ambitions ... we might even speak of a "race to the top"'(Albrecht et al. 2005, 177). Thus the data show upward convergence, toward more stringent measures, not only within the EU but also across the sample.

This does not mean, however, that the EU had no effect on the environmental policies of its member states. The ENVIPOLCON results indicate that the EU had an effect on raising standards, independent of the general trend towards convergence. In particular, the effect of EU membership was strong in the 1970s (Albrecht et al. 2005, 160). This is true for existing member states, not just states newly joining the EU. However, the trend reversed in the 1990s, with EU membership producing downward pressure on the level of standards. The researchers speculated:

> [t]his development might be the result of an overall change in governance patterns of EU environmental policy that is characterized by a shift from detailed command–and–control regulations towards procedural and non-detailed policies that leave member states much more leeway for compliance (Albrecht et al. 2005, 161).

This is consistent with predictions above about the impact of the shift away from quantitative standards, towards the more flexible approach of the Integrated Pollution Prevention and Control Directive.

Comparison with the Findings from the Classic Federations

How does the European Union experience compare to the American, Swiss and Canadian cases? There are some commonalities in terms of competitiveness concerns, the role of fiscal transfers and preferences for centralization. The EU differs from the other cases in its decision–making approach. Unlike the US approach, which currently requires no formal approval from state governments in standard setting, member states continue to play a strong role in EU environmental policy, although individual member states rarely have a veto, after the 1987 change to QMV. However, while intergovernmental cooperation is required in the EU process, the cooperation is not informal as in the Canadian case. Nor is there any real possibility of costless 'exit' from a directive once it has been passed.

As in the United States and Switzerland, a major impetus for centralized standard setting is the wish to create a 'level playing field', because of competitiveness concerns. As noted above, the justification of 'levelling conditions of competition' has been extended past the logical breaking point, to policy issues in which the internal market plays no role. Furthermore, in the development of EU directives, there is evidence of 'leader' member states seeking to 'export' their own stringent environmental standards to the remainder of the EU. For example, the 1988 Large Combustion Plants Directive 88/609 imposed limits on sulphur dioxide, nitrogen oxides and particulates. This directive was very similar in purpose and design to Germany's Large Combustion Plant Regulation of 1982. The German government sought to '… safeguard the competitiveness of domestic industry, which was now subject to more stringent requirements than its foreign competitors…' (Héritier et al. 1996, 179).

There is also evidence that competitiveness concerns have also impeded standard setting for pollution from industrial sources. Although framework directives on water and air pollution from industrial sources were passed relatively early on (in 1976 and 1984), it proved very difficult to pass the 'daughter directives' giving force to the policy by setting specific emissions limits. It has proven to be easier to set strict emissions limits for publicly owned pollution sources, such as waste water treatment plants or municipal solid waste incinerators.

As in the US and Switzerland, improvements in urban wastewater treatment appear to have been greatly facilitated by fiscal transfers from the center. It is very unlikely that the Urban Waste Water Treatment directive could have passed without guarantees of subsidies for implementation. Greece, Portugal and Spain made their approval of the directive conditional on the provision of subsidies from the Structural and Cohesion Funds, which are earmarked for regional development (Kemp 2002, 22).

As in Switzerland, publics across member states appear to have a strong preference for centralized environmental policymaking. While publics in member states are not in agreement over their preferences about environmental protection, they show substantial agreement about which level should make environmental policy. In 1996, 65 per cent of Europeans polled thought that environmental policy should be made at the EU level and 32 per cent thought it should be made at the national level (Eurobarometer 1997, B32). This is somewhat ironic because the principal of subsidiarity was first introduced into EU law solely in the context of environmental regulations. In contrast, only 54 per cent of those polled favoured a common EU currency. Only four other policy areas showed higher levels of support for EU decision–making. Unfortunately, Eurobarometer does not ask respondents why they favour decisionmaking at the EU level rather than the national level.

How do the findings about the EU differ from those in the other cases? Although all the cases have experience with unanimous decision–making in cooperative governance, the results of those decision–making processes differ. For example, in its preference for consensus on environmental policy, the Council of Ministers might seem similar to the Canadian Council of Ministers of the Environment (CCME). However, whereas the European Union has been able to advance an ambitious environmental policy program, Canada's interprovincial efforts have yielded modest results.

Decision–making by unanimity in the Council of Ministers did not prove an insuperable impediment to setting stringent standards. Many environmental directives were passed prior to the 1987 shift to QMV, including some of the most intrusive and expensive environmental directives ever passed. These include:

- the 1975 Bathing water directive 76/160;
- the 1976 framework water pollution directive 76/464, regulating emission of hazardous substances into water;
- the 1980 Drinking water directive 80/778;
- the 1984 Directive 84/360 of 28 June 1984, on the combating of air pollution from industrial plants.

Stringent directives have continued to be passed after the shift to QMV, such as the Urban Waste Water Treatment Directive and the directives on Municipal Solid Waste incineration. The challenge of passing stringent measures is likely to increase given the large number of poor countries which have joined the EU since accession has expanded into Eastern Europe.

None of the efforts at cooperative standard setting in the US, Switzerland or Canada can approximate this record. These either failed in their efforts to set common standards, or their standards were not stringent and not binding, as is the case with the Canadian harmonization process under the Canadian Council of Ministers of the Environment (CCME). With their explicit standards, deadlines for implementation and penalties for noncompliance, the EU directives regulations have far more in common with post–centralization legislation in the United States than CCME guidelines, which are more similar to measures found in international environmental agreements. Measures are voluntary and there are no penalties for noncompliance.

Why do EU rules look like national rules, while compacts or CCME guidelines look like international agreements? What explains the difference in these cases of intergovernmental bargaining? The difference cannot be explained due to the decision rule. For example, an account of the CCME's attempts at cooperative environmental policymaking attributes the weakness of these measures to the unanimity decision rule (Fafard 1998). However, prior to 1987, the Council of Ministers employed decision–making by unanimity, as did the cooperative processes in the three federal states. Thus the differences cannot simply be attributed to the decision rule. The exercise of vetoes can impede stringent standard setting but the existence of a veto does not rule out stringent standards.

Nor can enforceability be the only difference. As noted earlier, both American states and Swiss cantons have the capacity to write legally enforceable contracts. In theory, if states or cantons were concerned about defection from cooperation, they could create legal enforcement provisions within their compacts/concordats. They chose not to, however. The observations from the EU suggest that pre–existing legal institutions, exogenous to the standards, matter. EU member states are often able to negotiate concessions within a directive, but once a directive is passed, they cannot opt out of the directive. Nor can they opt out of enforcement provisions. In contrast, none of the standards agreed to by the CCME are binding upon provincial governments, even though they have all been negotiated by those governments, on a consensus basis. For environmental standard setting, there is, in practice, no equivalence exists between having recourse to enforcement and existing mechanisms for enforcement, exogenous to the agreement itself, which automatically apply to the agreement.

The key difference may be that EU directives, once passed are binding on all members. Although member states have been able to negotiate special concessions ('derogations') in the text of the directive, the terms of the enforcement mechanism are not negotiable. Nor can member states costlessly cannot 'opt out' of a directive once it enters into force.

Conclusion

The findings for the European Union are consistent with the pattern that a shift to standard setting at a higher level leads to more stringent standards. EU environmental policymaking, particularly in the early period, resulted in more stringent standards, not a lowest common denominator of member state standards. It is unclear whether this pattern will continue. Accession of new member states continues to make the European Union less homogenous, particularly with regard to wealth and level of development. Although EU standards will probably continue to be higher, on average, than what would have been the case without the EU, it may be difficult for the EU to continue as a global leader in environmental standards. The effects of the EU institutions and the heterogeneous preferences of its membership counteract each other.

As in any jurisdiction, concerns about competitiveness are always a factor in making environmental policy. Although these concerns are present, the structure of the European Union assuages those fears sufficiently to make strong environmental policy possible. Because decision–making in the Union is both intergovernmental and centralized, two patterns are observed. First, national governments are extremely wary of any measures which they feel might put their national industries at a competitive disadvantage. The British have been particularly vocal on this point. Second, the existence of an order of government above the national level has made it possible for countries to set more stringent standards, than they would have unilaterally. Also, high standard countries have successfully exported their standards to the remainder of the Union, in order to minimize their own competitive disadvantage *vis–à–vis* other member states.

In the earliest period of European Union environmental policymaking, the European Union did pass many stringent environmental regulations. The stringent policies were passed even though decisions were made unanimously in the Council of Ministers during this period. The lesson here is that, by itself, decision–making under unanimity does not preclude setting stringent, binding standards. Thus the fact that compacts, *Konkordaten* and the Canadian Council of Ministers of the Environment and international organizations all make decisions unanimously is not the sole reason for the weakness of the measures these groups create.

The salient difference between the European Union and cooperative decision–making bodies are the institutions of the European Union. The EU's institutional structure limits possibilities for exit from negotiations, because the membership is fixed, not ad hoc. The existence of the European Court of Justice as an enforcement mechanism appears to reduce the likelihood of defection from EU directives.

Chapter 9
Conclusion

Decentralized governance has been in favour on both sides of the Atlantic in recent years. Once advocacy of 'states' rights' lost its racist stigma, American conservatives began to tout the benefits of interjurisdictional completion, calling it 'environmental federalism'. Europeans on the other hand, have promoted the principle of subsidiarity and argued for the advantages of fluid and decentralized Multi–level Governance.

The cases examined here suggest that scepticism about these claims is warranted in the domain of environmental regulation, and perhaps other areas of prudential regulation. Environmental regulation presents a collective action problem, one best resolved by a centralized response. When environmental governance in these federations was most decentralized, lower levels of government proved unable to prevent serious environmental problems from growing worse. Strenuous efforts at cooperative solutions to transboundary pollution problems were similarly ineffective.

The chapter begins with a summary of the findings for each of the cases. The second section presents comparative data on rates of domestic wastewater treatment, which permit comparisons in environmental performance over time within cases and also across cases. The third section compares the findings from the cases with hypotheses about decentralization and cooperation. The fourth section identifies areas for future research.

The United States

Decentralized environmental governance, without federal floor standards, was not effective in the United States. Air and water quality in the US deteriorated steadily and this decline was not halted or reversed until federal legislation set minimum national standards in the late 1960s. Industry groups favoured state level standard setting and opposed federal legislation on air and water pollution. Concerns about create a level playing field in regulations were a significant motivator in the creation of federal standards for air and water pollution.

The US case also shows that, even when governments have the opportunity to create binding, enforceable contracts to address transboundary problems, they will not create them. The US Constitution authorizes interstate compacts as a way of solving interstate disputes, without federal intervention. Although interstate compacts have been used to address water pollution for 70 years, relatively few compacts were negotiated. The provisions of the compacts were generally weak and none provided for penalties for noncompliance. As a result, very few of the pollution control compacts were effective: only three compacts can be shown to have had *any* effect on reducing water pollution. Even here, the independent contribution of the compacts was limited because the biggest improvement was due to increased construction of municipal wastewater treatment facilities, facilities that were heavily subsidized by the US federal government, going as far back as the New Deal of the 1930s (US NRC 1939, 1).

Switzerland

Switzerland shows very high levels of environmental concern and has strong institutional and constitutional barriers to centralization. Environmental policy, once exclusively the responsibility of the cantons, has become more centralized since the 1950s. Most cantons did very little to protect the environment, despite evidence of serious water pollution as early the 1870s. The cantons, individually and collectively, were unable to provide the level of environmental protection that the majority of the Swiss population, across all cantons, sought. The cantons did not negotiate any formal treaties amongst themselves on water pollution prior to federal pre–emption of this issue. Under informal cooperation in Switzerland, cantonal water officials were eventually able to reach agreement on limited water pollution standards. This agreement was, however, subject to the condition that the standards they agreed upon be enshrined in federal legislation. As in the US case, the spread of sewage treatment was greatly facilitated by federal subsidies.

The steadily growing centralization of environmental policy has largely been driven by referenda which reassigned jurisdiction over a series of environmental issues from the cantons to the federal government. The significance of the referenda is that they set a very high hurdle for passage: a majority of all voters, but also majorities in a majority of cantons. Two of the most far reaching environmental referenda in 1953 and 1971 passed in every single canton. Thus Swiss in all cantons were clearly concerned about environmental conditions, yet cantonal governments remained unable to deliver effective environmental policies. After centralizing referenda passed, environmental quality in Switzerland began to improve dramatically.

Canada

Like Switzerland, Canada is a very noncentralized federation in many policy areas. In accordance with constitutional jurisprudence and informal political understandings, provincial governments have primary jurisdiction over environmental protection and thus tremendous autonomy in creating environmental policy. Federal intervention, through legislation or the courts, is fairly uncommon. Opinion polls show that Canadians are at least as concerned about the environment as Americans, yet an average Canadian factory pollutes at least twice as much as one in the US, and produces more pollution per job (Commission for Environmental Cooperation 1997).

Despite existing national legal institutions and well–developed fiscal arrangements for side payments, the Canadian provinces and the federal government have been unable to agree to binding environmental floor standards. Cooperative national standard setting was first attempted in 1975, with little success. The most recent effort – an accord on national environmental standards – did not contain any goals against which progress could be measured. While the Canadian Council of Ministers of Environment does publish guidelines for environmental quality, these are not binding, even though they have been agreed to by consensus of all provinces. Interprovincial environmental cooperation looks extremely similar to international environmental treaties: often vague and not enforceable. Canada's experience suggests that, even with high levels of public concern, effective environmental protection is very difficult to achieve cooperatively or in a very noncentralized context.

Comparative Data on Wastewater Treatment

Rates of municipal wastewater treatment are a good metric for comparing environmental performance across jurisdictions. First, these data control for differences in economic base, because municipal wastewater treatment is an issue for any agglomeration of people. Comparing pollution regulations for the steel industry, for example, would be comparing apples and oranges because not all jurisdictions have a steel industry. Second, studying wastewater treatment minimizes the nettlesome issue of enforcement. Comparative data on implementation and enforcement of environmental regulations are rarely available. This uncertainty is reduced when studying municipal wastewater treatment because it is easy to determine whether or not a wastewater treatment plant has been constructed and the level of treatment it provides. Once a plant has been constructed, there are few incentives for diverting wastewater around a wastewater treatment plant.

The data show that all countries have achieved significant improvements in their rates of wastewater treatment. European Union data are not presented here because the EU only regulated on this issue in 1991. The Netherlands and Western Germany are shown to provide a comparison with other wealthy countries with strong support for the environment. Switzerland shows a particularly dramatic rate of improvement. In 1930, a neighbourhood in Zurich was the only place in the country with a sewage treatment plant. In the early 1960s, the Swiss Federal government began to subsidize the construction of municipal sewage treatment plants.

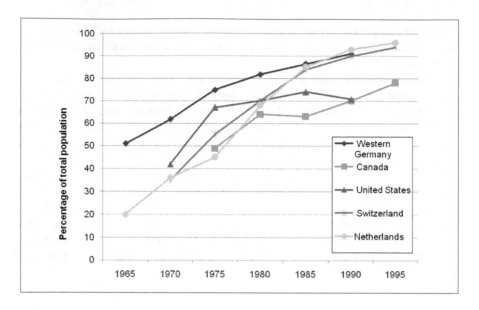

Figure 9.1 Per cent population connected to STP
Source: OECD 1979, 163; OECD 1997b, 74.

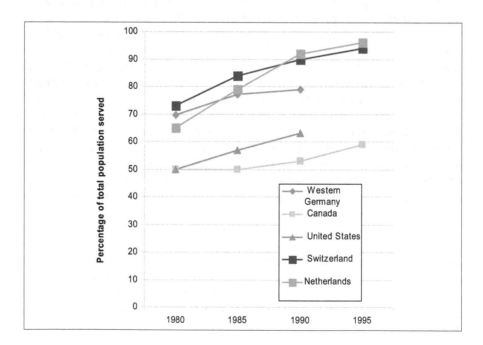

Figure 9.2 Second and tertiary sewage treatment
Source: OECD 1997b, 74.

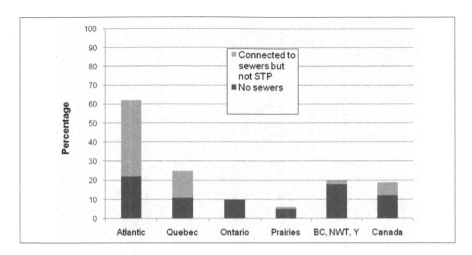

Figure 9.3 Canadian population not connected to a sewage treatment plant

Source: Chambers et al. 1997, 662.

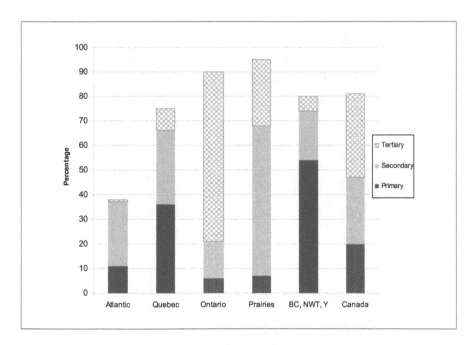

Figure 9.4 Canadians served by sewage treatment by type and region (1994)

Source: Chambers et al. 1997, 662.

Table 9.1 Hypotheses and findings from the cases

Hypothesis	Prediction	Findings	Conclusions
H₁: National governments are no better able to make environmental policy than subnational governments.	There should be no systematic relationship between centralization and the level of environmental protection.	As the US, Switzerland and the EU centralized environmental policymaking, the level of environmental protection increased. Canada lags significantly behind the US in most areas of environmental protection.	As a system becomes more centralized, the level of environmental protection increases.
H₂: Environmental policymaking by lower levels of jurisdiction is more likely to be tailored to local environmental conditions than centrally set policies.	Environmental policies set at lower levels of jurisdiction should show diversity, reflecting diversity in local conditions and problems.	Decentralized systems show great variation but the variation does not reflect the severity of local environmental problems. The majority of jurisdictions will have inadequate pollution control policies. In Canada and Switzerland, many places with a demonstrable need for domestic wastewater treatment did not (or do not) have it.	In the absence of minimum national standards, a wide range of policies is observed. However, even the most stringent jurisdictions tend to be less stringent than policies made at the national level. Many jurisdictions will have very limited environmental policies.
H₃: Lower levels of jurisdiction better reflect voters' preferences.	If lower levels of jurisdiction better reflect voter preferences, voters will prefer decentralized to centralized environmental policy.	In Switzerland, electorate has repeatedly voted, by wide margins, in favour of centralizing environmental policy in a series of referenda. European Union opinion polls show consistent support for making environmental policy at the EU, not national level.	People may be dissatisfied with the level of environmental protection provided under decentralization. This reflects a collective action problem.

H4: In the presence of favourable contracting conditions to deter defection, subnational governments will achieve effective cooperation on environmental problems.	In federal systems with high levels of concern about the environment, we should find many examples of effective cooperation.	Subnational cooperation on the environment looks like international cooperation. Agreements are limited, vague and unenforceable. Despite Swiss and American constitutional provisions permitting binding interstate agreement, there are few cases of such treaties for pollution control. The Swiss have never employed this legal mechanism. In the US, compacts were used in a dozen cases of water pollution but only three compacts can be shown to have had an independent effect on reducing pollution. The EU, setting standards cooperatively, has had far greater success.	Favourable contracting conditions are not a sufficient condition for effective environmental cooperation. Institutions matter. Fear of defection may not be the only obstacle to effective cooperation. Distributional issues among compact signatories may be an obstacle.

Figure 9.1, above suggests that all these countries are converging on a high level of treatment. The graph appears to suggest that Canada and the US have similar levels of municipal wastewater treatment. Neither impression is wholly accurate.

Figure 9.2 shows that there are significant disparities in the quality of treatment provided. The graph shows the percentage of the population whose sewage receives secondary or tertiary treatment. Secondary treatment is the minimum standard of treatment required of cities by the US federal government and the European Union. Canadian rates of secondary and tertiary treatment are low in comparative terms.

Figure 9.1 suggests little disparity between Canada and the US in rates of sewage treatment. These figures do not tell the whole story. They do not distinguish between populations not connected to sewers (such as those with septic tanks) and populations connected to sewers but not to sewage treatment plants. Because the Canadian population is more urban than the US population, a larger proportion of those not connected to sewage treatment are connected to city sewers. This is the least favourable outcome from an environmental point of view. In a sparsely populated area, septic tanks are an appropriate solution to managing sewage problems.

From a public health and water pollution standpoint, having populations connected to public sewers but not sewage treatment is far from ideal. In the US, only a small (and declining) percentage of the US population is in this situation. As of 1988, only 0.5 per cent of Americans (1.5 million) were connected to a sewer system that routinely (not accidentally) discharged untreated sewage (US EPA 1990, 148). In contrast, as of 1994 seven per cent of Canadians are connected to public sewers but not to any sewage treatment plant (Chambers et al. 1997, 662).

The percentage of those connected to sewers but not sewage treatment varies dramatically across Canada's regions from a high of 40 per cent in Atlantic Canada to a low of 0.1 per cent in Ontario. Due to improvements in Quebec, the percentage of Quebeckers connected to sewers but not treatment plants fell to 12 per cent by the mid 1990s (Chambers et al. 1997, 662). (When combined, the percentages in Figures 9.3 and 9.4 total 100 per cent. For example, Figure 9.3 shows 62 per cent of the population in Atlantic Canada have no sewage treatment. Figure 9.4 shows whether the remaining 38 per cent have primary, secondary or tertiary sewage treatment).

The Canadian case strongly suggests that decentralization matters for policy outcomes. The existence of very large disparities in sewage treatment, disparities which have persisted for decades, gives the lie to the notion that policies inevitably diffuse. It is not the case that decentralized systems reach the same endpoint as centralized systems.

Hypotheses and Findings from the Cases

Table 9.1 summarizes the findings about the hypotheses from the cases and predictions derived from them. Other things being equal, the stringency of a country's environmental policies is associated with the centralization of authority over environmental policy. Hence, within a country, environmental policy decided solely at the state level will be more lax than that decided at the national level, and in turn will be more stringent than environmental policy decided solely at the local level. In all cases, transitions to more centralized governance have been accompanied by more stringent environmental standards.

Tailoring to Local Conditions

Making environmental policymaking at lower levels of jurisdiction does not mean that those policies will be optimally adapted to local conditions. None of the cases presented here found evidence of environmental policies being tailored to local conditions under decentralization. Nor does centralized standard setting mean a one–size–fits–all approach. Due to lack of expertise and resources, as well as concerns about competitiveness, many jurisdictions will do nothing on most questions of environmental policy. Thus, to some extent, decentralization results in a uniform response of doing nothing.

Second, most centrally set standards are minima, which permit adaptation by local jurisdictions, provided the minimum standard is met. William R. Lowry found that state environmental policies did not closely match the severity of that state's environmental problems. In particular, policies were not tailored to local conditions if interstate economic competition was present and US federal government did not have a role in setting floor standards (Lowry 1992, 126).

For decentralized standard setting to lead to optimal standards, local knowledge must be more accurate than national level information. There is, however, a paradox in knowledge of local conditions. For people to know how much mercury is in their local fish, water or soil, they have to have access to data collected and analyzed by scientists. In addition, it is not enough to know what the mercury level is; some reference to a benchmark is also required. On average, lower levels of government lack the technical personnel or the resources to conduct such analyses. For example, European Union citizens only obtained accurate data on bacterial counts at local beaches once the EU Bathing Waters directive mandated the collection and publication of this information. There are economies of scale in environmental quality data. Noncentralized systems usually have weak capacity for monitoring and disseminating information on environmental quality. Thus, in most cases, individuals have little information about many local environmental conditions, particularly in a noncentralized system.

Representation of Preferences

In environmental policy, if jurisdictions are too small, the preferences of the majority *will not* be expressed in policy. The jurisdiction may be too small because of physical spillovers of pollution into and out of the jurisdiction, which reduce the incentive for unilateral action on pollution control. The more transboundary pollution there is, the less likely a jurisdiction is to set pollution abatement standards. However, it is just as likely that the jurisdiction is too small to overcome the collective action problems that arise because of concerns about competitiveness. Under either one of these conditions, decentralized decision–making does not maximize the number of people whose preferences are satisfied.

The Swiss have longstanding concern about environmental issues. When environmental policy was made by the cantons alone, Switzerland experienced deteriorating environmental quality. In 1971, a referendum authorized a constitutional amendment that gave the Swiss federal government a greater role in environmental policy. Ninety–two per cent of voters voted in favour, and in all cantons, the majority voted 'yes'. The results of this, and other Swiss referenda on the environment, indicate strong, broad based support for environmental protection yet these preferences were not reflected in the policies of the cantons.

What Might Explain the Discrepancy between National and Subnational Government Responses?

The above hypotheses do not propose mechanisms to explain why centralized environmental policymaking is more likely to be stringent and effective than noncentralized action by subnational governments. The following hypotheses suggest some mechanisms which could account for the systematic discrepancy between standards set locally and centrally.

Lack of Competence or Capacity at Lower Levels

The discrepancy could be accounted for by incompetence and lack of institutional capacity. Historical accounts of American state government behavior focused on lower levels of professionalism and lack of financial resources at the state level. This chararcterization is, however, less true now than ever. Furthermore, the claim of lesser competence does not hold in Canada, where provincial governments have been well–financed and professional for many decades. Thus, lesser competence does not seem useful as an explanation of the tendency of subnational governments to have less stringent environmental policies than the federal government.

Information Asymmetry

The lower the level of jurisdiction, the greater the information asymmetry between firms and government. Lower levels of government are at a disadvantage in collecting and analyzing data, particularly data useful in determining the credibility of threats of exit or impending bankruptcy. Smaller governments are less able to collect and assess industry–level data on costs, as well as data on regulatory requirements in other jurisdictions. Larger jurisdictions benefit from economies of scale in information collection and analysis. There are also economies of scale in expertise and learning by doing. A higher level of government is more likely to repeated experiences with firms in a particular sector, such as an auto plant or an aluminum smelter.

Risk and Redistribution

Larger entities are better able to spread risk. A town that acts unilaterally to control pollution from its largest employer runs the risk of cutting its own throat. If a large employer threatens exit, a larger jurisdiction can better afford to call the firm's bluff. If the firm actually leaves, a larger jurisdiction has more resources to redistribute to compensate the town which is affected. This phenomenon may be analogous to negotiating trade liberalization. The promise of compensation has made it possible to secure the acquiescence of labour and other groups, to more liberalized trade. If one looks at a smaller geographic territory with a correspondingly narrow economic base (such as textiles in South Carolina), one can clearly identify potential losers but it may be harder to imagine potential beneficiaries.

Systematic Differences in the Representation of Interest at Different Levels of Jurisdiction

The relative leverage of business and environmental groups may be more equal at the national level, rather than at the local level. Because environmental groups have fewer resources than many firms, they can make the most effective use of their research and financial resources by focusing their efforts at the national level. Businesses might also lobby more effectively, than environmental groups, at lower levels of government because the disparity between concentrated costs and diffuse benefits is greater at lower levels of jurisdiction.

If environmental groups mobilize their constituencies through national media and advertising, and not grassroots canvassing or neighbourhood activism, they can be expected to have more impact at the national level than locally. In practice, testing this hypothesis would be complicated by the need to test for regulatory capture as a competing hypothesis which could produce the same results. One differentiating feature might be that regulatory capture would probably vary across state governments, by industry sector.

Greater Ability to Capture Rents at Lower Levels of Government

The tendency for lower levels of government to enact less stringent environmental protection may have nothing whatsoever to do with competition. Rather, it may be easier for organized groups to extract rents from lower levels of government than from the federal government. Two of the groups which have most effectively avoided environmental regulation have little credible threat of exit: agriculture and services provided by small businesses. These groups argue that environmental protection will put them out of business but they cannot realistically claim that they will locate to another jurisdiction.

Lower levels of government may be more susceptible to capture because some industries are so dominant in a particular jurisdiction. When businesses and industries lobby intensely at the federal level, their demands may cancel out. For example, American sugarcane growers want to protect sugar quotas but the processed food industry has a substantial interest in free trade in sugar. At lower levels of jurisdiction, there may be fewer countervailing interests. For example, in Florida, sugar cane growers will have far more clout than jam makers. Environmental groups are likely to be at a disadvantage with regard to industry but there may also be fewer competing industry groups to protest special treatment for one industry.

Conditions for Effective Cooperation

Some cooperative institutions have been effective in protecting the environment. However, the efficacy of such measures or noncentralized policy can be increased by centralized authority, or the threat of centralization of authority. Minimum standards imposed by the centre are more conducive to state level innovation than a totally noncentralized system. In addition, if cooperative institutions operate in the shadow of hierarchy, participants may be motivated to work harder at cooperation, if failure may result in pre–emption by the centre. Some of the more effective water pollution control compacts in the US were negotiated in the shadow of several trips to the Supreme Court.

Subnational governments are only rarely able to solve environmental problems cooperatively. The US had seven decades of experience with cooperative water quality management. Only three interstate compacts demonstrably improved water quality of the river in question. Surprisingly, cooperation on water pollution control was not easier to achieve on lakes than on rivers.

Before the creation of the Environmental Protection Agency (EPA), the federal government sought to promote environmental protection by encouraging interstate cooperation within watersheds and airsheds. The federal government provided grants for infrastructure and for training but interstate cooperation remained woefully inadequate. The EPA pre–empted state authority in this area beginning with the *Clean Water Act* in 1965 and since then, the US has witnessed dramatic improvements in air and water quality.

From the examples in federal systems explored in this book, it appears that the criteria of concern, capacity and contracting conditions are not sufficient for effective environmental cooperation. Most of the cases in this book show high levels of concern and have institutions for contract enforcement or extensive experience with interstate or federal/provincial agreements. In addition, it is difficult to argue that the states or provinces studied in these cases lack the *capacity* to make environmental policy. These same governments have responsibility for massive government programs and complex types of regulation.

Although ineffective forms of environmental cooperation employ unanimous decision–making, the evidence from the European Union suggests that a consensus decision–making rule does not necessarily lead to lax standards or ineffective measures. The Council of Ministers generally prefers to make decisions by consensus even though the formal rule for most environmental regulations now is Qualified Majority Voting (QMV). It is significant that many of the EU's most stringent and demanding policies were passed unanimously, before the advent of QMV in the Council of Ministers.

Even though states governments in a federation have recourse to external enforcement mechanisms, they may not avail themselves of it. This has been the case with American interstate compacts and Swiss *Konkordats*. American interstate compacts are enforceable by federal courts. Those creating interstate environmental compacts could have drafted them in such a way to make them enforceable in federal court. But state governments chose not to do so. Having recourse to enforcement mechanisms outside the contract is of little effect if recourse is not used.

The success of EU environmental policymaking relative to interstate cooperation suggests that the existence of recourse to enforcement is not equivalent to integrated institutions for enforcement. It appears that the transaction costs involved in cooperative environmental policymaking are very substantial. Cooperative and hierarchical governance structures are not equivalent. It seems to make a difference that the enforcement structure and the terms of enforcement are exogenous to the agreement to be enforced. The European Union and hierarchical governance also appear to have an advantage over cooperative governance in their ability to supply side–payments. This capacity to offer side–payments helps overcome distributional issues that can be a barrier to agreement.

Areas for Future Research

This book identifies two major areas for future research. First, all the cases presented here offer evidence that industries consistently prefer standard setting by lower levels of jurisdiction. Similarly, in all the cases, environmentalists favour standard setting by higher levels of government. How generalizable are these results? Secondly, there is a need for additional comparative research on the relationship between governance structures and efficacy, in particular to isolate the impact of specific institutional design characteristics.

There is a need for comparative studies of lobbying at different levels of government. Do industries have a systematic advantage over environmentalists when lobbying at the subnational level? Why do environmental groups appear to have a greater impact at the national than the subnational level? It is possible that subnational governments are more susceptible to industry capture but it is not clear why this should be the case. There are many cases of regulatory capture at the US federal level of government.

Secondly, studies of the beliefs of policymakers, elected officials and bureaucrats, about regulation and competitiveness are also needed. Are governments as fearful of industry exit as they appear to be? Do bureaucrats and politicians at the national level have different beliefs than those at the subnational level? And, the greatest mystery of all, why are decision makers so fearful of economic impact of environmental regulations when economists offer no support for these views? This issue may be analogous to commonly held mercantilist views, where few politicians appear to believe in the benefits of free trade. Politicians are generally in favour of exports but do not appear to agree with economists that under free trade, imports are beneficial as well.

Comparative study of the relationship between governance structures and efficacy would yield valuable insights to help design better institutions. Comparisons between international level governance and nation states do not yield much insight because there is insufficient variation on key institutional variables. There is a need for comparative study of governance models showing variation on features such as membership, voting rules, monitoring and enforcement. This is particularly important for creating more effective international institutions, where fewer features of design are manipulable.

References

Adler, R., Landman, J. and Cameron, D. (1993), *The Clean Water Act 20 Years Later* (Washington DC: Island Press).

Ahmad, E. and Tanzi, V. (eds.) (2002), *Managing Fiscal Decentralization* (New York: Routledge).

Albrecht, J., Arts, B. and Liefferink, D. (2005), 'The Gap Approach to Policy Convergence', in *Environmental Governance in Europe: TShe Impact of International Institutions and Trade on Policy Convergence: Quantitative Report.* <http://www.uni–konstanz.de/FuF/Verwiss/knill/projekte/envipolcon/downloads/QuantRep_2005.pdf>

'Aluminum Smelter Costs'. (2000), *Mining Journal*, 17 March, 209.

A. J. Chandler & Associates. (2006), *Review of Dioxins and Furans from Incineration: In Support of a Canada-wide Standard Review.* <http://www.ccme.ca/assets/pdf/1395_d_f_review_chandler_e.pdf>

Anderson, T. and Hill, P. (eds.) (1997), *Environmental Federalism* (Lanham: Rowman & Littlefield).

Andrews, R. (1999), *Managing the Environment, Managing Ourselves: A History of American Environmental Policy* (New Haven: Yale University Press).

Armingeon, K. (1998), 'Renegotiating the Swiss Welfare State', in Van Waarden and Lehmbruch (eds.).

—— (2000), 'Swiss Federalism in Comparative Perspective', in Wachendorfer–Schmidt (ed.).

Arp, H. (2005), 'Technical Regulation and Politics: the Interplay between Economic Interests and Environmental Policy Goals in EC Car Emission Legislation', in Jordan (ed.).

Ashby, E. and Anderson, M. (1981), *The Politics of Clean Air* (Oxford: Clarendon Press).

Axelrod, R. (1984), *The Evolution of Cooperation* (New York: Basic Books).

—— and Keohane, R. (1986), 'Achieving Cooperation under Anarchy: Strategies and Institutions', in Oye (ed.).

Bache, I. and Flinders, M. (eds.) (2004), *Multi-level Governance* (Oxford: Oxford University Press).

Bakker, K. (ed.) (2007), *Eau Canada: The Future of Canada's Water* (Vancouver: UBC Press).

Ball, S. and Bell, S. (1991), *Environmental Law* (London: Blackstone Press).

Banting, K. (1982), *The Welfare State and Canadian Federalism* (Kingston: McGill–Queen's University Press).

Barbera, A. and McConnell, V. (1990), 'The Impact of Environmental Regulations on Industry Productivity: Direct and Indirect Effects', *Journal of Environmental Economics and Management* 18:1, 50–65.

Barrett, S. (1991), 'The Problem of Global Environmental Protection', in Helm (ed.).

Bartik, T. (1985), 'Business Location Decisions in the U.S.: Estimates of the Effect of Unionization, Taxes and Other Characteristics of States', *Journal of Business and Economic Statistics* 3:1, 14–22.

——— (1988), 'The Effects of Environmental Regulation on Business Location United States', *Growth and Change* 19:3, 22–44.

——— (1989), 'Small Business Start–ups in the United States: Estimates of the Effects of Characteristics of States', *Southern Economic Journal* 55:4, 1004–8.

Bauer, T. and Davidson, E. (1989), *Silbersonne am Horizont: ALUSUISSE—Eine Schweizer Kolonialgeschichte* (Zürich: Limmat Verlag).

Baumol, W. and Oates, W. (1988), *The Theory of Environmental Policy* (New York: Cambridge University Press).

Baxter, J. (1999), 'Government Can't Enforce Environment Law: Report', *The Ottawa Citizen*, 7 September , A5.

Baxter, L. (1974), *Regional Politics and the Challenge of Environmental Planning.* (Davis: Institute of Governmental Affairs, University of California).

Becker, C. and Neitzel, D. (eds.) (1992), *Water Quality in North American River Systems* (Columbus: Battelle Press).

Becker, R. and Henderson, V. (2000), 'Effects of Air Quality Regulations on Polluting Industries', *Journal of Political Economy* 108:2, 379–421.

Begg, D. et al. (1993), *Making Sense of Subsidiarity: How Much Centralization for Europe?* (London: Centre for Economic Policy Research).

Benidickson, J. (2007), *The Culture of Flushing: A Social and Legal History of Sewage* (Vancouver: UBC Press).

Bernauer, T. (1996), 'Protecting the Rhine River against Chloride Pollution', in Keohane and Levy (eds.).

Beyle, T. (ed.) (1989), *State Government* (Washington DC: Congressional Quarterly).

Biersteker, T. and Weber, C. (eds.). (1996), *State Sovereignty as a Social Construct* (Cambridge, UK: Cambridge University Press).

Bhagwati, J. and Hudec, R. (eds.) (1996), *Fair Trade and Harmonization: Prerequisites for Free Trade?* (Cambridge, MA: MIT Press).

Bird, P. and Rapport, D. (1986), *State of Environment Report for Canada* (Ottawa: Ministry of Supply and Services).

Blazejczak, J. and Löbbe, K. (1993), *Umweltschutz und Industriestandort : der Einfluss umweltbezogener Standortfaktoren auf Investitionsentscheidungen* (Berlin: Erich Schmidt).

Blendon, R., Benson, J., Morin, R., Altman, D., Brody, M., Brossard, M. and James, M. (1997), 'Changing Attitudes in America', in Nye et al. (eds.).

Blom–Hansen, J. (1999), 'Avoiding the "Joint Decision Trap"': Lessons from Intergovernmental Cooperation', *European Journal of Political Research* 35:1, 35–67.

Boardman, R. (ed.) (1992), *Canadian Environmental Policy: Ecosystems, Politics and Process* (Toronto: Oxford University Press).

Bochsler, D. and Sciarini, P. (2006), 'Konkordate und Regierungskonferenzen: Standbeine des Horizontalen Föderalismus', *Leges* 1.

Bohn, G. (1989), 'Study Reveals That Majority of Pulp Mills Break Law: Worst Polluters in Canada Found in BC and Quebec', *Vancouver Sun*, 15 March, A1.

Bolleyer, N. (2006a), 'Consociationalism and Intergovernmental Relations – Linking Internal and External Power–Sharing in the Swiss Federal Polity', *Swiss Political Science Review* 12:3, 1–34.

—— (2006b), 'The Internal Life of Subnational Governments, Interdepartmental Spill–over and the Organizational Convergence of Intergovernmental Relations', paper given at the DVPW Kongress, Münster Germany, 26 September 2007.

—— (2007), 'Internal Government Dynamics and the Nature of Intergovernmental Relations' (Ph.D. thesis, European University Institute).

Bonner, M. (1994), *The National Sewage Report Card: Rating the Treatment Methods and Discharges of 20 Canadian Cities* (Vancouver: Sierra Legal Defense Fund).

Börzel, T. (2003), *Environmental Leaders and Laggards in Europe: Why There Is (Not) a 'Southern Problem'* (Aldershot: Ashgate Publishing).

Boyd, D. (2007), *Prescription for a Healthy Canada: Towards a National Environmental Health Strategy* (Vancouver: David Suzuki Foundation).

Bozinoff, L. and MacIntosh, P. (1989), 'Canadians Increasingly Concerned about Dangers of Pollution', *Gallup* (Toronto: Gallup Canada Inc). 19 June 1989.

Brandenburger, K., Gasche, U., Guggenbühl, H. and Haemmerli, F. (1982), *Das Märchen von der sauberen Schweiz: Pleite im Umweltschutz* (Basel: Lenos Verlag).

Brautigam, T. (2008), 'NL Struggles to Close "Industrial Age" Incinerators', 26 October. < http://cnews.canoe.ca/CNEWS/Politics/2008/10/26/7211476-cp.html>

Breeze, L. (1993), *The British Experience with River Pollution, 1865–1876* (New York: Peter Lang).

Brennan, G. and Buchanan, J. (1980), *The Power to Tax: Analytical Foundations of the Fiscal Constitution* (New York: Cambridge University Press).

Breton, A. (1991), 'The Existence and Stability of Interjurisdictional Competition', in Kenyon and Kincaid (eds.).

—— (1996), *Competitive Governments: an Economic Theory of Politics and Public Finance* (New York: Cambridge University Press).

—— (2002), 'An Introduction to Decentralization Failure', in Ahmad and Tanzi (eds.).

Brown, D. (2002), *Market Rules: Economic Union Reform and Intergovernmental Policy–Making in Australia and Canada* (Montreal: McGill–Queen's University Press).

Buchanan, J. (1962), 'Politics, Policy and Pigouvian Margins', *Economica* 29, 17–28.

—— (1968), *The Demand and Supply of Public Goods* (Chicago: Rand McNally).

—— (1995), 'Federalism as an Ideal Political Order and an Objective for Constitutional Reform', *Publius* 25:2, 19–27.

—— and Tullock, G. (1975), 'Polluters' Profits and Political Response: Direct Controls vs. Taxes', *American Economic Review* 65:1, 139–47.

Buchholz, D. (1998), 'Competition and Corporate Incentives: Dilemmas in Economic Development' (Ph.D. dissertation, Department of Political Science, Duke University).

Burki, S., Perry, G. and Dillinger, W. (1999), *Beyond the Center: Decentralizing the State* (Washington DC: The World Bank).

Burton, Jr., J. (1995), '"Workers" Compensation Costs: Regional Variations in 1995', *Workers' Compensation Monitor* 8:6.

Bussmann, W. (1981), *Gewässerschutz und kooperativer Föderalismus in der Schweiz* (Bern: Verlag Paul Haupt).

—— (1986), *Mythos und Wirklichkeit der Zusammenarbeit im Bundesstaat: Patentrezept oder Sackgasse?* (Bern: Verlag Paul Haupt).

Butler, D. and Ranney, A. (eds.) (1994), *Referendums around the World: The Growing Use of Direct Democracy*. Washington DC: AEI Press).

Butler, H. and Macey, J. (1996), *Using Federalism to Improve Environmental Policy* (Washington DC: AEI Press).

Byatt, I. (1996), 'The Impact of EC Directives on Water Customers in England and Wales', *Journal of European Public Policy* 3:4, 665–74.

Calzonetti, F. and Walker, R. (1991), 'Factors Affecting Industrial Location Decisions: a Survey Approach', in Herzog, Jr. and Schlottmann (eds.).

Canada. (1991), *The State of Canada's Environment* (Ottawa: Minister of Supply and Services).

Canada. House of Commons. Standing Committee on Environment and Sustainable Development. (1997), *Harmonization and Environmental Protection: An Analysis of the Harmonization Initiative of the Canadian Council of Ministers of the Environment* (Ottawa: House of Commons). <http://cmte.parl.gc.ca/cmte/CommitteePublication.aspx?SourceId=36163>

—— (1998), *Enforcing Canada's Pollution Laws*. <http://cmte.parl.gc.ca/cmte/CommitteePublication.aspx?COM=110&Lang=1&SourceId=36189>

'Canada "Outspews" the US: Average Plant Here Pollutes More Than Twice as Much' (1997), *Ottawa Citizen*, 29 July, A1.

Canadian Council of Ministers of the Environment (CCME). (1994), *Annual Report '93 – '94* (Winnipeg: CCME).

—— (1998), 'A Canada–wide Accord on Environmental Harmonization', <http://www.ccme.ca/assets/pdf/accord_harmonization_e.pdf>

—— (1999), 'Socio–Economic Analysis for Dioxins and Furans: Summary by Priority Sector' (Winnipeg: CCME).

—— (2000a), 'Environment Ministers Meet at Kananaskis', <http://www.ccme.ca/about/communiques/2000.html?item=4>

—— (2000b), 'Canada–wide Standards for Dioxins and Furans', 6 June 2000.

—— (2000c), 'Initial Set of Actions. Dioxins and Furans Canada–wide Standards. Emissions from Incinerators and Coastal Pulp and Paper Boilers', 6 June 2000.

—— (2007), *Canada–wide Strategy for the Management of Municipal Wastewater Effluent: DRAFT*. Winnipeg MB: CCME. <http://www.ccme.ca/assets/pdf/mwwe_cda_wide_strategy_consultation_e.pdf>

Canadian Opinion Research Archive. Queen's University. <http://www.queensu.cora/_trends/ap_Environment.htm>

Carment, D., Hampson, F. and Hillmer, N. (eds.) (2003), *Canada among Nations 2003: Coping with the American Colossus* (Don Mills ON: Oxford University Press).

Carroll, J., Hyde, M. and Hudson, W. (1987), 'State–Level Perspectives on Industrial Policy: the Views of Legislators and Bureaucrats', *Economic Development Quarterly* 1:4, 333–42.

Cary, W. (1974), 'Federalism and Corporate Law: Reflections upon Delaware', *Yale Law Journal* 83:4, 663–705.

Cassell, E. (1968), 'The Health Effects of Air Pollution and Their Implications for Control', *Law and Contemporary Problems* 33:2, 197–216.

Cassils, J. (1991), *Exploring Incentives: An Introduction to Incentives and Economic Instruments for Sustainable Development* (Ottawa: National Round Table on the Environment and the Economy).

Castleman, B. (1978), 'How We Export Dangerous Industries', *Business and Society Review* 27: 7–14.

—— (1979), 'The Export of Hazardous Factories to Developing Nations', *International Journal of Health Services* 9:4, 569–606.

Centre for Research on Direct Democracy. <http://www.c2d.ch/inner.php?table=votes

Chadwick, E. (1965), *Report on the Sanitary Condition of the Labouring Population of Great Britain* (1843, repr. Edinburgh: Edinburgh University Press).

Chambers, P. (1969), 'Water Pollution Control through Interstate Agreement', *University of California Davis Law Review* 1: 43–70.

Chambers, P. et al. (1997), 'Impacts of Municipal Wastewater Effluents on Canadian Waters: A Review', *Water Quality Research Journal of Canada* 32:4, 659–713.

Chayes, A. and Chayes, A. (1995), *The New Sovereignty: Compliance with International Regulatory Agreements* (Cambridge, MA: Harvard University Press).

City of Ottawa. (2007), 'Ottawa 20/20: Infrastructure Masterplan', <http://www.
    ottawa.ca/city_services/planningzoning/2020/imp/annex1_en.shtml>

Clark, J. (1938), *The Rise of a New Federalism: Federal–State Cooperation in the
    United States* (New York: Columbia University Press).

Clark, K. and Winfield, M. (1996),'The Environmental Management Framework
    Agreement – A Model for Dysfunctional Federalism?' CIELAP Brief 96/1.
    (Toronto: Canadian Institute for Environmental Law and Policy).

Clarke, H. and Stewart, M. (1997), 'Green Worlds and Public Deeds: Environmental
    Hazards and Citizen Response in Canada and the United States', in Frizzell
    and Pammett (eds.).

Cloghesy, M. (1998), 'The Benefits of Harmonizing our Environmental Regulatory
    System', *CanadaWatch* 6:1.

Coase, R. (1937), 'The Nature of the Firm', *Economica* 4, 386–405.

—— (1960), 'The Problem of Social Cost', *Journal of Law and Economics* 3,
    1–44.

Colodey, A. and Wells, P. (1992), 'Effects of Pulp and Paper Mill Effluents on
    Estuarine and Marine Ecosystems in Canada: A Review', *Journal of Aquatic
    Ecosystem Health* 1:3, 201–26.

Collier, U. (ed.) (1998), *Deregulation in the European Union: Environmental
    Perspectives* (London: Routledge).

Commission for Environmental Cooperation (1997), *Taking Stock North American
    Pollutant Releases and Transfers* (Montreal: Commission for Environmental
    Cooperation).

'Committee Okays Cleansing Project'. (2008), *Montreal Gazette*, 24 April, A7.

Congleton, R. (ed.) (1996), *The Political Economy of Environmental Protection:
    Analysis and Evidence* (Ann Arbor: University of Michigan Press).

Conybeare, J. (1980), 'International Organization and the Theory of Property
    Rights', *International Organization* 34:3, 307–34.

Cooter, R. (1982), 'The Cost of Coase', *Journal of Legal Studies* 11:1, 1–33.

Cornes, R. and Sandler, T. (1986), *The Theory of Externalities, Public Goods and
    Club Goods* (New York: Cambridge University Press).

Council of the Federation (2008), 'Trade: Building on our Strengths in Canada
    and Abroad', <http://www.councilofthefederation.ca/pdfs/COMMUNIQUE_
    TRADE_clean.pdf>

Council of State Governments (1956), *Interstate Compacts 1783–1956* (Chicago:
    Council of State Governments).

Crandall, R. (1993), *Manufacturing on the Move* (Washington DC: The Brookings
    Institution).

Cropper, M. and Oates, W. (1992), 'Environmental Economics: The Survey',
    Journal of Economic Literature 30:2, 675–740.

Curlin, J. (1972), 'The Interstate Water Pollution Compact: Paper Tiger or Effective
    Regulatory Device?' *Ecology Law Quarterly* 2: 333–56.

Dahl, R. (1967), 'The City in the Future of Democracy', *American Political
    Science Review* 61:4, 953–70.

—— (1994), 'A Democratic Dilemma: System Effectiveness vs. Citizen Participation', *Political Science Quarterly* 109:1, 23–34.

Dales, J. (1968), *Pollution, Property and Prices* (Toronto: University of Toronto Press).

David, P. and Eisenberg, R. (1961), *Devaluation of the Urban and Suburban Vote* (Charlottesville: Bureau of Public Administration, University of Virginia).

David Suzuki Foundation. (2006), *The Water We Drink: An International Comparison of Drinking Water Standards and Guidelines* (Vancouver: David Suzuki Foundation). <http://www.davidsuzuki.org/files/SWAG/DSF–HEHC–water–web.pdf>

Davies, C. (1969), *The Politics of Pollution* (New York: Pegasus).

—— (1975), *The Politics of Pollution*, 2nd Edition (Indianapolis: Pegasus).

Dawson, R. and Robinson, J. (1963), 'Inter–party Competition, Economic Variables and Welfare Policies in the American States', *Journal of Politics* 25:2, 265–89.

Dean, J. (1992), 'Trade and the Environment: A Survey of the Literature', in Low (ed.).

Degler, S. (1970), *State Air Pollution Control Laws* (Washington DC: Bureau of National Affairs).

Deily, M. and Gray, W. (1991), 'Enforcement of Pollution Regulations in a Declining Industry', *Journal of Environmental Economics & Management* 21:3, 260–74.

Deschamps, Y. (1977), <http://en.wikipedia.org/wiki/Yvon_Deschamps#Quotes>

Deudney, D. (1995), 'The Philadelphian System: Sovereignty, Arms Control and Balance of Power in the American States–union, circa 1787–1861', *International Organization* 49:2, 191–228.

—— (1996), 'Binding Sovereigns: Authorities, Structures and Geopolitics in Philadelphian Systems', in Biersteker and Weber (eds.).

Dimock, M. and Benson, G. (1937), *Can Interstate Compacts Succeed?* (Chicago: University of Chicago Press).

Dingle, A. (1982), '"The Monster Nuisance of All": Landowners, Alkali Manufacturers and Air Pollution 1828–64', *Economic History Review* 35:4, 529–48.

Dodd, A. (1936), 'Interstate Compacts', *United States Law Review* 70: 141–166.

Doern, G. and Johnson, R. (eds.) (2006), *Rules, Rules, Rules, Rules: Multilevel Regulatory Governance* (Toronto: University of Toronto Press).

—— and MacDonald, M. (1999), *Free–Trade Federalism: Negotiating the Canadian Agreement on Internal Trade* (Toronto: University of Toronto Press).

Donahue, J. (1997), *Disunited States* (New York: Basic Books).

Dose, N., Holznagel, B. and Weber, V. (1994), *Beschleunigung von Genehmigungsverfahren: Vorschlage zur Verbesserung des Industriestandortes Deutschland* (Bonn: Economica Verlag).

Downs, G., Rocke, D. and Barsoom, P. (1996), 'Is the Good News about Compliance Good News about Cooperation?' *International Organization* 50:3, 379–406.

Duerksen, C. (1983), *Environmental Regulation of Industrial Plants Citing: How to Make It Work Better* (Washington DC: The Conservation Foundation).

Duffy–Deno, K. (1992), 'Pollution Abatement Expenditures and Regional Manufacturing Activities', *Journal of Regional Science* 32:4, 419–36.

Dyck, R. (1970), 'Evolution of Federal Air Pollution Control Policy' (Ph.D. dissertation, Environmental Sciences, University of Pittsburgh).

Dye, T. (1968), *Politics, Economics and the Public* (Chicago: Rand McNally).

—— (1990), *American Federalism: Competition among Governments* (Lexington: Lexington Books).

EAWAG. (1981), *Gewässerschutz in der Schweiz: Sind die Ziele erreichbar?* (Bern: Verlag Paul Haupt).

Economic Council of Canada (1981), *Reforming Regulation* (Ottawa: The Council).

Eisler, P. (1999a), 'Data Flawed on Drinking Water Quality', *USA Today* 2 September 2 1999, 1A.

—— (1999b), 'Water Monitors May Be "Flying Blind"', *USA Today,* 2 September 1999, 5A.

Elazar, D. (ed.) (1991), *Federal Systems of the World* (New York: Stockton).

Engel, K. (1997), 'State Environmental Standard–Setting: Is There a "Race" and Is It "to the Bottom"?' *Hastings Law Journal* 48:2, 271–376.

Environment Canada. CEPA Environmental Registry. <http://www.ec.gc.ca/ CEPARegistry/agreements/Eqv_Agree.cfm>

—— (2007a), *Canada's 2005 Greenhouse Gas Inventory: A Summary of Trends.* <www.ec.gc.ca/pdb/ghg/inventory_report/2005/2005summary_e.cfm>

—— (2007b), 'Government of Canada Takes Action to Combat Dumping of Raw Sewage and Upgrade Sewage Treatment', 24 September 2007. <http://www. ec.gc.ca/default.asp?lang=En&n=714D9AAE–1&news=FA83843D–2731– 4CA1–8750–C75F380B572A>

—— , Statistics Canada and Health Canada (2007), *Canadian Environmental Sustainability Indicators: Highlights.* Catalogue no. 16–252–XIE, <www. statscan.ca>

Erikson, R., Wright, G. and MacIver, J. (1993), *Statehouse Democracy: Public Opinion and Policy in the American States* (New York: Cambridge University Press).

Eurobarometer (1997), *Eurobarometer 46* <http://ec.europa.eu/public_opinion/ archives/eb/eb46/tab46.pdf>

European Commission. DG Environment. 2001. 'The Challenge of Environmental Financing in the Candidate Countries'. COM (2001)304.

EEA (European Environment Agency) (2005), 'Effectiveness of Urban Wastewater Treatment Policies in Selected Countries: An EEA Pilot Study'.(Copenhagen: European Environment Agency). <http://reports.eea.europa.eu/eea_report_ 2005_2/en/FINAL_2_05_Waste_water_Web.pdf>

Evans, P. (1999), 'Japan's Green Aid Plan: The Limits of State–Led Technology Transfer', *Asian Survey* 39:6, 825–44.

Fafard, P. (1998), 'Green Harmonization: the Success and Failure of Recent Environmental Intergovernmental Relations', in Lazar (ed.).

—— and Harrison, K. (eds.) (2000), *Managing the Environmental Union* (Montreal: McGill–Queen's University Press).

Farnsworth, S. (1999), 'Federal Frustration, State Satisfaction? Voters and Decentralized Governmental Power', *Publius* 29:3, 75–88.

Farrell, J. (1987), 'Information and the Coase Theorem', *Journal of Economic Perspectives* 1:2, 113–29.

Featherstone, K. and Radaelli, C. (eds.) (2003), *The Politics of Europeanization* (New York: Oxford University Press).

Feld, L., Pommerehne, W. and Hart, A. (1996), 'Private Provision of the Public Good: a Case Study', in Congleton (ed.).

Filippov, M., Ordeshook, P. and Shvetsova, O. (2004), *Designing Federalism: A Theory of Self-sustainable Federal Institutions.* (New York: Cambridge University Press).

Fine, S. (1997), 'Ontario among Top Polluters: Province Follows Two US States as Worst Offenders in North America, According to Study', *Globe and Mail,* 29 July, A1, A4.

Fite, E. (1932), *Government by Cooperation* (New York: Macmillan).

Flick, C. (1980), 'The Movement for Smoke Abatement in Nineteenth–Century Britain', *Technology and Culture* 21:1, 29–50.

Flinn, M. (1965), 'Introduction' in E. Chadwick, *Report on the Sanitary Condition of the Labouring Population of Great Britain.* 1842; reprint, (Edinburgh: Edinburgh University Press).

Flynn, B. (2000), 'Is Local Truly Better? Some Reflections on Sharing Environmental Policy between Local Governments and the EU', *European Environment* 10, 75–84.

Frankfurter, F. and Landis, J. (1925), 'The Compact Clause of the Constitution—A Study in Interstate Adjustments', *Yale Law Journal* 34:7, 685–758.

Franson, M., Franson, R. and Lucas, A. (1982), *Environmental Standards: a Comparative Study of Canadian Standards, Standard Setting Processes and Enforcement* (Edmonton: Environment Council of Alberta).

Frenkel, M. (1986), 'Interkantonale Institutionene und Politikbereiche', in Germann and Weibel (eds.).

Friedman, J., Gerlowski, D. and Silberman, J. (1992), 'What Attracts Foreign Multinational Corporations? Evidence from Branch Plant Location in the United States', *Journal of Regional Science* 32:4, 403–18.

Frizzell, A. and Pammett, J. (eds.) (1997), *Shades of Green: Environmental Attitudes in Canada and Around the World.* International Social Survey Programme (ISSP) Series #2. (Ottawa: Carleton University Press).

Gallup Institute. (1972), *The Gallup Poll: Public Opinion 1935-1971, vol. 1 1935–1949* (NY: Random House).

'Garbage Incinerators to Get Extension, Minister Says' (2008), 15 October. <http://www.cbc.ca/canada/newfoundland-labrador/story/2008/10/15/teepee-extensions.html>

Gasche, U. (1981), *Bauern, Klosterfrauen, Alusuisse* (Gümligen: Zytglogge Verlag).

Gaspari, K. and Woolf, A. (1985), 'Income, Public Works and Mortality in Early Twentieth-Century American Cities', *Journal of Economic History* 45:2, 355–61.

Gerbely, F. (1989), 'Kanton Alusuisse: Alusuisse im Wallis', in Bauer and Davidson.

Germann, R. and Weibel, E. (eds.) (1986), *Handbuch Politisches System der Schweiz. Band 3: Föderalismus* (Bern: Paul Haupt Verlag).

Giroux, L. (1987), 'Delegation of Administration', in Tingley (ed.).

Golub, J. (1996), 'British Sovereignty and the Development of EC Environmental Policy', *Environmental Politics* 5:4, 700–28.

Goodstein, E. (1999), *The Trade–off Myth: Fact and Fiction about Jobs and the Environment* (Washington DC: Island Press).

Graham Jr., F. (1966), *Disaster by Default: Politics and Water Pollution* (New York: M. Evans and Co.).

Grandy, C. (1989), 'New Jersey Corporate Chartermongering, 1875–1929', *Journal of Economic History* 49:3, 677–92.

Grant, W. (1998), 'Large Firms, SMEs and Deregulation', in Collier (ed.).

—— , Matthews, D. and Newell, P. (2000), *The Effectiveness of European Union Environmental Policy* (New York: St. Martin's Press).

Gray, W. (1997), 'Manufacturing Plant Location: Does State Pollution Regulation Matter?', National Bureau of Economic Research Working Paper no. 5880.

Greenstein, F. and Polsby, N. (eds.) (1975), *Handbook of Political Science, vol. 5.* (Reading, MA: Addison–Wesley Publishing Co).

Grodzins, M. (1966), *The American System: A New View of Government in the United States* (Chicago: Rand McNally).

Haas, P., Keohane, R., and Levy, M. (eds.) (1993), *Institutions for the Earth: Sources of Effective International Environmental Protection* (Cambridge, MA: MIT Press).

Hahn, P. (2006), 'A Conversation with the Prime Minister', 23 December 2006. <http://www.ctv.ca/servlet/ArticleNews/story/CTVNews/20061221/harper_year_end_061221/20061223?hub=TopStories>

Hall, J. (ed.) (1997), *Air Pollution and Regional Economic Performance: a Case Study* (Greenwich, CT: JAI Press).

Hanf, K. and Scharpf, F. (eds.) (1978), *Interorganizational Policy Making: Limits to Coordination and Central Control* (Beverly Hills: Sage Publications).

Harrington, J. and Warf, B. (1995), *Industrial Location: Principles, Practice and Policy* (London: Routledge).

Harrington, W., Morgenstern, R. and Nelson. P. (2000), 'On the Accuracy of Regulatory Cost Estimates', *Journal of Policy Analysis and Management* 19:2, 297–322.

Harrison, K. (1995a), 'Is Co-operation the Answer? Canadian Environmental Enforcement in Comparative Context', *Journal of Policy Analysis and Management* 14:2, 221–44.

—— (1995b), 'Federalism, Environmental Protection and Blame Avoidance', in Rocher and Smith (eds.).

—— (1996a), 'The Regulator's Dilemma: Regulation of Pulp Mill Effluents in the Canadian Federal State', *Canadian Journal of Political Science* 29:3, 469–96.

—— (1996b), *Passing the Buck: Federalism and Canadian Environmental Policy* (Vancouver: UBC Press).

—— (ed.) (2006), *Racing to the Bottom? Provincial Interdependence in the Canadian Federation* (Vancouver: UBC Press).

Hastings, E. and Hastings, P. (1982), *Index to International Public Opinion 1980–1981* (Westport CT: Greenwood Press).

Haveman, R. and Christiansen, G. (1981), 'Environmental Regulations and Productivity Growth', in Peskin et al. (eds.).

Haverland, M. (2003), 'The Impact of the European Union on Environmental Policies', in Featherstone and Radaelli (eds.).

Hayek, F. (1978), 'Competition as a Discovery Procedure', in F. Hayek, *New Studies in Philosophy, Politics, Economics, and the History of Ideas* (Chicago: University of Chicago Press).

Heerings, H. (1993), 'The Role of Environmental Policies in Influencing Patterns of Investments of Transnational Corporations: Case Study of the Phosphate Fertilizer Industry', in OECD.

Helm, D. (ed.) (1991), *Economic Policy toward the Environment* (Cambridge, UK: Blackwell Publishers).

Henderson, J. (1996), 'Effects of Air Quality Regulation', *American Economic Review* 86:4, 789–813.

Héritier, A., Knill, C. and Mingers, S. (1996), *Ringing the Changes in Europe: Regulatory Competition and the Transformation of the State* (New York: Walter de Gruyter).

Herzog Jr., H. and Schlottmann, A. (eds.) (1991), *Industry Location and Public Policy* (Knoxville: University of Tennessee Press).

Heustis, L. (1984), *Policing Pollution: The Prosecution of Environmental Offences* (Ottawa: Law Reform Commission of Canada).

Hill, C. (2006), 'Two Models of Multi-level Governance, One Model of Multi-level Accountability' (Ph.D. thesis, University of British Columbia).

—— and Harrison, K. (2006), 'Intergovernmental Regulation and Municipal Drinking Water', in Doern and Johnson (eds.).

——, Furlong, K., Bakker, K. and Cohen, A. (2007), 'A Survey of Water Governance Legislation and Policies in the Provinces and Territories', in Bakker (ed.).

Hillyard, D. (1995), 'From Tomorrow to Today: The Canadian Council of Ministers of the Environment and Canadian Environmental Policy' (M.A. thesis, University of Windsor).

Hoberg, G. (1992), 'Comparing Canadian Performance in Environmental Policy', in Boardman (ed.).

Holmes, T. (1998), 'The Effect of State Policies on the Location of Manufacturing: Evidence from State Borders', *Journal of Political Economy* 106:4, 667–705.

Holzinger, K. (1994), *Politik des kleinsten gemeinsamen Nenners?* (Berlin: Edition Sigma).

——, Knill, C. and Sommerer, T. (2008), 'Environmental Policy Convergence: The Impact of International Harmonization, Transnational Communication and Regulatory Competition', *International Organization* 62: 4, 553–87.

Hooghe, L. and Marks, G. (2003), 'Unraveling the Central State, but How? Types of Multi-level Governance', *American Political Science Review* 97: 233–43.

Hunt, H. (1997), 'Discussion', *New England Economic Review* (March/April).

Indermaur, P. (1989), 'Was andere können, können wir auch: Eine Geschicthe Alusuisse', in Bauer and Davidson.

Inglehart, R., Basanez, M. and Moreno, A. (1998), *Human Values and Beliefs: A Cross–Cultural Sourcebook* (Ann Arbor: University of Michigan Press).

Inman, R. and Rubenfeld, D. (1997), 'Rethinking Federalism', *Journal of Economic Perspectives* 11:4, 43–64.

Jacob, H. and Lipsky, M. (1968), 'Outputs, Structures and Power: An Assessment of Changes in the Study of State and Local Politics', *Journal of Politics* 30: 510–38.

Jacobs, D. (1999), 'How Industry Beat the Environmental Protection Act: After an Intense and Prolonged Lobbying Campaign the Draft EPA has been Declawed', *Ottawa Citizen*, 7 September 1999, A1.

Jaffe, A., Peterson, S., Portney, P. and Stavins, R. (1995), 'Environmental Regulation and the Competitiveness of U.S. Manufacturing: What Does the Evidence Tell Us?' *Journal of Economic Literature* 33:1, 132–163.

Jans, J. (1996), 'Legal Protection in European Environmental Law: An Overview', in Somsen (ed.).

Jewell, M. (1989), 'What Hath *Baker v. Carr* Wrought?', in Beyle (ed.).

Johnston, R. (1986), *Public Opinion and Public Policy in Canada: Questions of Confidence* (Toronto: University of Toronto Press).

Jones, C. (1975), *Clean Air: the Policies and Politics of Pollution Control* (Pittsburgh: University of Pittsburgh Press).

Jones, K. and Bucher, J. (1998), 'Will Reducing Transported Ozone Improve Regulatory Compliance?' *EM* (July): 11–16.

Jordan, A. (2002), *The Europeanization of British Environmental Policy* (New York: Palgrave Macmillan).

—— (ed.) (2005), *Environmental Policy in the European Union*, 2nd Edition (London: Earthscan Publications).

Kahn, M. (1997), 'Particulate Pollution Trends in the United States', *Regional Science and Urban Economics* 27:1, 87–107.

—— (1999), 'The Silver Lining of Rust Belt Manufacturing Decline', *Journal of Urban Economics* 46:3, 360–76.

Katz, R. (1998), 'Malapportionment and Gerrymandering in Other Countries and Alternative Electoral Systems', in Rush (ed.).

Keating, T. and Farrell, A. (1999), 'Transboundary Environmental Assessment: Lessons from the Ozone Transport Assessment Group', Technical Report NCEDR/99–02. (Knoxville: National Center for Environmental Decision-Making Research).

Kehoe, T. (1997), *Cleaning up the Great Lakes: From Cooperation to Confrontation* (Dekalb: Northern Illinois University Press).

Kemp, R. (2002), *Treatment of a EU Directive* (Maastricht: MERIT).

Kenyon, D. and Kincaid, J. (eds.) (1991), *Competition among States and Local Governments: Efficiency and Equity in American Federalism* (Washington DC: The Urban Institute Press).

Keohane, R. (1983), 'The Demand for International Regimes', in Krasner (ed.).

—— (1984), *After Hegemony: Cooperation and Discord in the World Political Economy* (Princeton: Princeton University Press).

—— and Levy, M. (1996), *Institutions for Environmental Aid: Pitfalls and Promise* (Cambridge, MA: MIT Press).

—— , Haas, P. and Levy, M. (1993), 'The Effectiveness of International Environmental Institutions', in Haas et al. (eds.).

Kerwin, K. and Grover, R. (1991), 'California Steamin': Business Makes Tracks from LA', *Business Week* 13 May 1991, 44–5.

Knill, C. and Liefferink, D. (2007), *Environmental Politics in the European Union.* New York: Manchester University Press).

Knödgen, G. (1979), 'Environment and Industrial Siting: Results of an Empirical Survey of Investment by West German Industry in Developing Countries', *Zeitschrift für Umweltpolitik* 2:4, 403–18.

Knox, R. (2000), 'The Unpleasant Reality of Interprovincial Trade Disputes', *Fraser Forum* October 2000.

Knutsen, H. (1999), 'Leather Tanning, Environmental Regulations and Competitiveness in Europe: a Comparative Study of Germany, Italy and Portugal', F·I·L Working Papers No. 17 (Oslo: Department of Human Geography, University of Oslo).

Kobach, K. (1994), 'Switzerland', in Butler and Ranney (eds.).

Koontz, T. (2002), *Federalism in the Forest: National versus State Natural Resource Policy* (Washington DC: Georgetown University Press).

Koppen, I. (1993), 'The Role of the European Court of Justice', in Liefferink et al. (eds.).

Krämer, L. (1998), *EC Treaty and Environmental Law*, 3rd Edition (London: Sweet & Maxwell).

—— (2007), *EC Environmental Law*, 6th Edition (London: Sweet & Maxwell).

Krasner, S. (ed.) (1983), *International Regimes* (Ithaca NY: Cornell University Press).

Laplante, B. and Rilstone, P. (1996), 'Environmental Inspections and Emissions of the Pulp and Paper Industry in Quebec', *Journal of Environmental Economics and Management* 33:1, 19–36.

Lazar, H. (ed.) (1998,) *Non–constitutional Renewal* (Kingston ON: Institute of Intergovernmental Relations).

Ledyard, J. (1984), 'The Pure Theory of Large Two–Candidate Elections', *Public Choice* 44:1, 7–41.

Leebron, D. (1996), 'Lying Down with Procrustes: An Analysis of Harmonization Claims', in Bhagwati and Hudec (eds.).

Leonard, H. (1988), *Pollution and the Struggle for the World Product* (Cambridge, UK: Cambridge University Press).

—— and Duerksen, C. (1980), 'Environmental Regulation in the Location of Industry: an International Perspective', *Columbia Journal of World Business* 15:2, 52–68.

Leuchtenburg, W. (1953), *Flood Control Politics: The Connecticut River Valley Problem 1927–1950* (Cambridge, MA: Harvard University Press).

Levinson, A. (1993), 'Studies in the Economics of Local Environmental Regulation', (Ph.D. thesis, Department of Economics, Columbia University).

—— (1996a), 'Environmental Regulations and Industry Location: International and Domestic Evidence', in Bhagwati and Hudec (eds.).

—— (1996b), 'Environmental Regulations and Manufacturers' Location Choices: Evidence from the Census of Manufactures', *Journal of Public Economics* 62,1–2, 5–29.

—— (1997), 'A Note on Environmental Federalism: Interpreting Some Contradictory Results', *Journal of Environmental Economics and Management* 33, 359–66.

Liefferink, D., Lowe, D. and Mol, A. (eds.) (1993), *European Integration & Environmental Policy* (New York: John Wiley and Sons).

Lijphart, A. (1997), 'Presidential Address 1996: Unequal Participation', *American Political Science Review* 91:1, 1–14.

Linder, W. (1998), *Swiss Democracy: Possible Solutions to Conflict in Multicultural Societies,* 2nd Edition (New York: St. Martin's Press).

Lipset, S. (1990), *Continental Divide: Values and Institutions of the United States and Canada* (New York: Routledge).

Llewellyn, K. (1934), 'The Constitution as an Institution', *Columbia Law Review* 34:1, 1–40.

Low, P. (ed.) (1992), *International Trade and the Environment.* World Bank Discussion Paper No. 159 (Washington DC: The World Bank).

—— and Yeats, A. (1992), 'Do 'Dirty' Industries Migrate?', in Low (ed.).

Lowry, W. (1992), *The Dimensions of Federalism: State Governments and Pollution Control Policies* (Durham, NC: Duke University Press).

Lucas, A. (1986), 'Harmonization of Federal and Provincial Environmental Policies: the Changing Legal and Policy Framework', in Saunders (ed.).

—— (1987), 'Natural Resource and Environmental Management: A Jurisdictional Primer', in Tingley (ed.).

—— (1989), 'The New Environmental Law', in Watts and Brown (eds.).

—— (1990), 'Jurisdictional Disputes: Is 'Equivalency' a Workable Solution?' in Tingley (ed.).

—— and Sharvit, C. (2000), 'Underlying Constraints on Intergovernmental Cooperation in Setting and Enforcing Environmental Standards', in Fafard and Harrison (eds.).

Lucas, R., Wheeler, D. and Hettige, H. (1992), 'Economic Development, Environmental Regulation and the International Migration of Toxic Industrial Pollution: 1960–1988', in Low (ed.).

Lyne, J. (1990), 'Service Taxes, International Site Selection and the 'Green' Movement Dominate Executives' Political Focus', *Site Selection*. October, 1134.

—— (1991), 'Anheuser–Busch's Jack Stein: Fusing Site Selection and Environmental Expertise', *Site Selection*. June, 586.

McConnell, V. and Schwab, R. (1990), 'The Impact of Environmental Regulation on Industry Location Decisions: The Motor Vehicle Industry', *Land Economics* 66: 67–81.

—— and Schwarz, G. (1992), 'The Supply and Demand for Pollution Control: Evidence from Wastewater Treatment', *Journal of Environmental Economics and Management* 23:1, 54–77.

Macdonald, D. (1991), *The Politics of Pollution: Why Canadians are Failing their Environment* (Toronto: McClelland & Stewart).

McInnes, C. (1988), 'Exemption from PCB Storage Rule Is Sought by All Provinces but PEI'. *Globe and Mail*, 14 Oct., A12.

McKinley, C. (1955), 'The Management of Water Resources under the American Federal Systems', in MacMahon (ed.).

MacMahon, A. (ed.) (1955), *Federalism: Mature and Emergent* (Garden City NY: Doubleday and Co).

Macrory, R. (ed.) (2005,) *Reflections on 30 years of EU Environmental Law: A High Level of Protection* (Groningen: Europa Law Publishing).

Mahtesian, C. (1994), 'Romancing the Smokestack', *Governing* November: 38.

Mäler, K-G. (1991), 'International Environmental Problems', in Helm (ed.).

Mann, D. (ed.) (1982), *Environmental Policy Implementation* (Lexington MA: Lexington Books).

Marks, G. and Hooghe, L. (2004), 'Contrasting Visions of Multi-Level Governance', in Bache and Flinders (eds.).

Markusen, J., Morey, E. and Olewiler, N. (1993), 'Environmental Policy When Market Structure and Plant Locations Are Endogenous', *Journal of Environmental Economics and Management* 24:1, 69–86.

—— (1995), 'Competition in Regional Environmental Policies When Locations Are Endogenous', *Journal of Public Economics* 56:1, 55–77.

Marotte, B. (1990), 'The Sudden Greening of the Pulp Makers: A Tide of Support for Tough Environmental Laws is Forcing the Industry to Clean Up its Act', *Financial Times of Canada*, 78:34, 7.

Mauch, C. and Reynard, E. (2002), *The Evolution of the National Water Regime in Switzerland* (Lausanne: IDHEAP).

Mitchell, C. (1995), 'Three Essays on Environmental Regulation' (Ph.D. dissertation, Department of Economics, University of California at Los Angeles).

Moore, J., Neel, C., Herzog Jr., C. and Pace, A. (1991), 'The Efficacy of Public Policy', in Herzog, Jr. and Schlottmann (eds.).

Moore, T. (1992), 'The Battle of Blackpool Beach', *Daily Telegraph* 28 October, 2.

Muldoon, P. and Valiante, M. (1989), 'Toxic Water Pollution in Canada: Regulatory Principles for Reduction and Elimination with Emphasis on Canadian Federal and Ontario Law' (Toronto: Canadian Institute for Environmental Law and Policy).

Murdoch, J., Sandler, T. and Sargent, K. (1997), 'A Tale of Two Collectives: Sulfur vs. Nitrogen Oxide Emissions Reduction in Europe', *Economica* 64, 281–301.

Murphy, D. (1995), 'Open Economies and Regulations: Convergence and Competition among Jurisdictions' (Ph.D. dissertation, Massachusetts Institute of Technology).

—— (2004), *The Structure of Regulatory Competition: Corporations and Public Policies in a Global Economy* (Oxford: Oxford University Press).

Musgrave, R. and Musgrave. P. (1989), *Public Finance in Theory and Practice* (New York: McGraw–Hill).

Muys, J. (1971), 'Interstate Water Compact: the Interstate Compact and the Federal Interstate Compact', Report to the National Water Commission (NWC-L-71-011).

National Round Table on the Environment and the Economy. (1996), *State of the Debate on the Environment and the Economy: Water and Wastewater Services in Canada* (Ottawa: National Round Table on the Environment and the Economy).

Nemetz, P. (1986), 'The Fisheries Act and Federal–Provincial Environmental Regulation: Duplication or Complementarity?' *Canadian Public Administration* 29:3, 401–24.

Neumayer, E. (2001), *Greening Trade and Investment: Environmental Protection without Protectionism* (London: Earthscan).

Nichols, M. and Jensen, H. (1990), 'Danger in the Water,' *Maclean's*, 15 January, 30–1.

'NLEIA approves of planned incinerator shutdown'. (2008), *The Northern Pen* (St. Anthony, NL) 29:19, A14. 5 May.

Nordell, J. (1999), 'Town Meeting Time', *Christian Science Monitor,* 18 March 1999, 12.

Nye, J., Zelikow, P. and King, D. (eds.) (1997), *Why People Don't Trust Government* (Cambridge, MA : Harvard University Press).

OECD (Organization for Economic Cooperation and Development) (1979), *The State of the Environment in OECD Member Countries* (Paris: OECD).

—— (1993), *Environmental Policies and Industrial Competitiveness* (Paris: OECD).

—— (1997a), *Environmental Policies & Employment* (Paris: OECD).

—— (1997b), *OECD Environmental Data* (Paris: OECD).

—— (2004a), *Environmental Performance Reviews: Canada* (Paris: OECD).

—— (2004b), 'OECD Review of Canada's Environmental Performance: Good Progress, Much to Be Done', <http://www.oecd.org/document/30/0,2340,en_2649_201185_33744542_1_1_1_1,00.html>

Oates, W. (1969), 'The Effect of Property Taxes and Local Public Spending on Property Values: an Empirical Study of Tax Capitalization and the Tiebout Hypothesis', *Journal of Political Economy* 77:6, 957–71.

—— (1972), *Fiscal Federalism* (New York: Harcourt Brace Jovanovich).

—— (1985), 'Searching for Leviathan: an Empirical Study', *American Economic Review* 75:4, 748–57.

—— and Schwab, R. (1988), 'Economic Competition among Jurisdictions: Efficiency Enhancing or Distortion Inducing?' *Journal of Public Economics* 35:3, 333–54.

Obinger, H. (1998), 'Federalism, Direct Democracy and Welfare State Development in Switzerland', *Journal of Public Policy* 18:3, 241–63.

Office of the Auditor General of Canada. (1999), 'Chapter 5. Streamlining Environmental Protection through Federal–Provincial Agreements: Are They Working?' *Report of the Commissioner of the Environment and Sustainable Development.* <http://www.oag-bvg.gc.ca/internet/English/parl_cesd_199905_e_1141.html>

—— (2005), 'Chapter 4. Safety of Drinking Water: Federal Responsibilities', *Report of the Commissioner of the Environment and Sustainable Development to the* House of Commons (Ottawa: Office of the Auditor General of Canada). <http://www.oag-bvg.gc.ca/internet/English/parl_cesd_200509_04_e_14951.html>

Olewiler, N. (2006), 'Environmental Policy in Canada: Harmonized at the Bottom?' in Harrison (ed.).

—— and Dawson, K. (1998), 'Analysis of national pollution release inventory data on toxic emissions by industry', Working Paper 97–16, Prepared for the Technical Committee on Business Taxation, Department of Finance, Government of Canada.

Ontario. (2000), 'Setting Environmental Quality Standards in Ontario: The Ministry of the Environment's Standards Plan'. <http://www.ene.gov.on.ca/envision/env_reg/er/documents/2000/pa9e0004.pdf>

Oye, K. (ed.) (1985), *Cooperation under Anarchy* (Princeton: Princeton University Press).

Pallemaerts, M. (2005), 'EC Chemicals Legislation: A Horizontal Perspective', in Macrory (ed.).

Parlour, J. (1981), 'The Politics of Water Pollution Control', *Journal of Environmental Management* 12:1, 31–64.

—— and Schatzow, S. (1978), 'The Mass Media and Public Concern for Environmental Problems in Canada, 1960–1972', *International Journal of Environmental Studies* 13:1, 9 –17.

Patrick, R. (1992), *Surface Water Quality: Have the Laws Been Successful?* (Princeton: Princeton University Press).

Patterson, J. (1969), *The New Deal and the States* (Princeton: Princeton University Press).

Pearson, C. (1987), *Multinational Corporations, Environment and the Third World* (Durham NC: Duke University Press).

Pérez-Peña, R. (1999a), 'Northeast States Softening Stance on Air Standards', *New York Times*, 28 August, A1.

—— (1999b), 'States Fail to Make Deal in Air Battle over Pollution', *New York Times*, 2 September, A16.

Peskin, H., Portney, P. and Kneese, A. (eds.) (1981), *Environmental Regulation and the U.S. Economy* (Washington DC: Resources for the Future).

Platt, R. (1980), *Intergovernmental Management of Floodplains*. Program on Technology, Environment and Man Monograph #30 (Boulder: Institute for Behavioral Science, University of Colorado).

Pomp, R. (1988), 'The Role of State Tax Incentives in Attracting and Retaining Businesses', *Economic Development Review* 6:2, 53–62.

Posse, A. (1986), *Föderative Politikverflechtung in der Umweltpolitik* (Munich: Minerva Publikation).

Prud'homme, R. (1995), 'The Dangers of Decentralization', *World Bank Research Observer* 10:2, 201–20.

Puri, A. (1997), 'A Survey Approach to Firm Location Decisions', in Hall (ed.).

Rabe, B. (1990), *Beyond NIMBY: Hazardous Wastes Siting in Canada and United States* (Washington DC: The Brookings Institution).

—— (1995), 'Integrating Environmental Regulation: Permitting Innovation at the State Level', *Journal of Policy Analysis and Management* 14:3, 467–72.

—— (1998), 'Federalism and Entrepreneurship: Explaining American and Canadian Innovation in Pollution Prevention and Regulatory Integration', *Policy Studies Journal* 27:2, 288–306.

Revesz, R. (1992), 'Rehabilitating Interstate Competition: Rethinking the "Race-to-the-Bottom" Rationale for Federal Environmental Regulation', *New York University Law Review* 67: 1210–54.

Revkin, A. (2005), 'US Resists New Targets for Curbing Emissions', *New York Times*, 8 Dec. http://www.nytimes.com/2005/12/08/science/earth/08climate.html

Ridker, R. and Henning, J. (1967), 'The Determinants of Residential Property Values with Special Reference to Air Pollution', *Review of Economics and Statistics* 49:2, 246–57.

Riechmann, V. (1978), *Die Vorbereitung bundeseinheitlicher gliedstaatlicher Gesetzgebung in den Vereinigten Staaten von Amerika als Problem des kooperativen Foederalismus* (Frankfurt am Main: Peter Lang).

Riker, W. (1975), 'Federalism', in Greenstein and Polsby (eds.).

Ringquist, E. (1993), *Environmental Protection at the State Level: Politics and Progress in Controlling Pollution* (Armonk NY: M.E. Sharpe).

Rocher, F. and Smith, M. (eds.) (1995), *New Trends in Canadian Federalism* (Peterborough ON: Broadview Press).

Rogers, P. (1996), *America's Water: Federal Roles and Responsibilities* (Cambridge, MA: MIT Press).

Rose–Ackerman, S. (1983), 'Beyond Tiebout: Modeling the Political Economy of Local Government', in Zodrow (ed.).

Rosen, H. (ed.) (1988), *Fiscal Federalism: Quantitative Studies* (Chicago: University of Chicago Press).

Rubin, E. and Feeley, M. (1994), 'Federalism: Some Notes on a National Neurosis', *UCLA Law Review* 41, 903–52.

Runnals, D. (1992), '1992 Environmental Scan Summary', Prepared for the Canadian Council of Ministers of the Environment by the Institute for Research on Public Policy.

Rush, M. (ed.) (1998), *Voting Rights and Redistricting in the United States* (Westport CT: Greenwood Press).

Sallot, J. (2005), 'Canada Lags on Air-pollution Cleanup Compared with US, Coalition Finds', *The Globe and Mail* 13 October, A7.

Sandel, M. (1995), *Democracy's Discontent: America in Search of a Public Philosophy* (Cambridge, MA: Harvard University Press).

Saunders, O. (ed.) (1986), *Managing Natural Resources in a Federal State* (Calgary: Canadian Institute of Resources Law).

Scarpino, P.(1985), *Great River: An Environmental History of the Upper Mississippi 1890 –1950* (Columbia: University of Missouri Press).

Scharpf, F. (1988), 'The Joint–Decision Trap: Lessons from German Federalism and European Integration', *Public Administration* 66:3, 239–78.

—— (1993), 'Introduction', in Scharpf (ed.).

—— (ed.) (1993), *Games in Hierarchies and Networks: Analytical and Empirical Approaches to the Study of Governance Institutions* (Boulder CO: Westview Press).

—— (2006), 'The Joint Decision Trap Revisited', *Journal of Common Market Studies* 44:4, 845–64.

Schenkel, W. (1998), *From Clean Air to Climate Policy in the Netherlands and Switzerland—Same Problems, Different Strategies?* (Bern: Peter Lang).

Schmalensee, R. and Willig, R. (eds.) (1989), *Handbook of Industrial Organization* (New York: North–Holland).

Schmulbach, J., Hesse, L. and Bush, J. (1992), 'The Missouri River — Great Plains Thread of Life', in Becker and Neitzel (eds.).

Sebenius, J. (1992), 'Challenging Conventional Explanations of International Cooperation: Negotiation Analysis and The Case Of Epistemic Communities', *International Organization* 46:1, 323–65.

Shannon, J. (1991), 'Federalism's "Invisible Regulator" — Interjurisdictional Competition', in Kenyon and Kincaid (eds.).

Shapiro, D. (1995), *Federalism: A Dialogue* (Evanston IL: Northwestern University Press).

Shrybman, S. (1988), *Environmental Impacts of Bill C–130: The Canada–US Free Trade Agreement as Environmental* Law (Toronto: Canadian Environmental Law Association).

Shubik, M. (1984), *A Game-Theoretic Approach to Political Economy* (Cambridge, MA: MIT Press).

Shugart , W. and Tollison, R. (1985), 'Corporate Chartering', *Economic Inquiry* 23:4, 585–99.

Sierra Legal Defence Fund. (2004), *The National Sewage Report Card: Grading the Sewage Treatment of 22 Canadian Cities (Number 3)* (Vancouver: Sierra Legal Defence Fund).

Sinclair, W. (1990), *Controlling Pollution from Canadian Pulp and Paper Manufacturers: A Federal Perspective* (Ottawa: Ministry of Supply and Services).

—— (1991), 'Controlling Effluent Discharges from Canadian Pulp and Paper Manufacturers', *Canadian Public Policy* 17:1, 86–105.

Smith, F. (1979), *The People's Health 1830–1910* (New York: Holmes & Meier Publishers).

Somsen, H. (ed.). (1996), *Protecting the European Environment: Enforcing EC Environmental Law* (London: Blackstone Press).

Sorsa, P. (1994), 'Competitiveness and Environmental Standards—Some Exploratory Results', Policy Research Working Paper no. 1249 (Washington DC: The World Bank).

Stafford, H. (1985), 'Environmental Protection and Industrial Location', *Annals of the Association of American Geographers* 75:2, 227–40.

Stern, A. (1982), 'History of Air–Pollution Legislation in the United States', *Journal of the Air Pollution Control Association* 32:1, 44–61.

Stiglitz, J. (1983), 'The Theory of Local Public Goods Twenty-Five Years after Tiebout: A Perspective', in Zodrow (ed.).

Strong, D. (1984), *Tahoe: An Environmental History* (Lincoln: University of Nebraska Press).

Sundquist, J. (1968), *Politics and Policy: The Eisenhower, Kennedy and Johnson Years* (Washington DC: The Brookings Institution).

Switzerland BUWAL (Federal Office of Environment, Forests and Landscape) (1994), *The State of the Environment in Switzerland* (Bern: BUWAL)

Sylves, R. (1982), 'Congress, EPA, the States and the Fight to Decentralize Water–Pollution–Grant Policy', in Mann (ed.).

Tannenwald, R. (1997), 'State Regulatory Policy and Economic Development', *New England Economic Review* (March/April): 83–98.

Tarr, J. (1996), *The Search for the Ultimate Sink: Urban Pollution in Historical Perspective* (Akron OH: University of Akron Press).

Taylor, M. (1987), *The Possibility of Cooperation* (Cambridge, MA: Cambridge University Press).

'The Spectator's View'. (1990), *The Daily Telegraph*, 1 June, 19.

Thompson, A. (1980), *Environmental Regulation in Canada: An Assessment of the Regulatory Process* (Vancouver: Westwater Research Center, University of British Columbia).

Tiebout, C. 1956. 'A Pure Theory of Local Expenditures', *Journal of Political Economy* 64:5, 416–24.

Tingley, D. (ed.) (1987), *Environmental Protection and Canadian Constitution* (Edmonton: Environmental Law Center).

—— (ed.) (1990), *Into the Future: Environmental Law and Policy for the 1990s* (Edmonton: Environmental Law Center).

Trebilcock, M., Prichard, J. and Courchene, T. (eds.) (1983), *Federalism and the Canadian Economic Union* (Toronto: University of Toronto Press).

UNCTAD/CTC. (1993), *Environmental Management in Transnational Corporations*. Report of the Benchmark Corporate Environmental Survey. Environment Series No. 4 ST/CTC 149. New York: United Nations.

United Nations Environment Programme. (1999), *Dioxin and Furan Inventories* (Geneva: UNEP Chemicals).

US. Advisory Commission on Intergovernmental Relations (1993), *Changing Public Attitudes on Governments and Taxes* (Washington DC: Superintendent of Documents).

US Congress. Senate. Committee on Public Works (1963), *A Study of Pollution— Air*, 88th Congress, 1st Session.

—— (1975), *Environmental Protection Affairs of the Ninety–Third Congress*, 94th Congress, 1st Session.

—— Subcommittee on Air and Water Pollution (1972), *Hearings on S. 907 a Bill to Consent to the Interstate Environment Compact.* 92nd Congress.

US CBO (Congressional Budget Office) (1985), *Environmental Regulation and Economic Efficiency* (Washington DC: United States Government Printing Office).

US EPA (Environmental Protection Agency). (1990), *National Water Quality Inventory: 1988 Report to Congress* (Washington DC: Office of Water).

US GAO (General Accounting Office). (1981), *Federal–Interstate Compact Commissions: Useful Mechanisms for Planning and Managing River Basin Operations.* (Washington DC: GAO).

—— (1986), *The Nation's Water: Key Unanswered Questions about the Quality of Rivers and Streams.* GAO/PEMD–86–6 (Washington DC: Superintendent of Documents).

US NRC (National Resources Committee). (1935), *Regional Factors in National Planning* (Washington DC: US Government Printing Office).

—— (1939), *Water Pollution in the United States. Third Report of the Special Advisory Committee on Water Pollution* (Washington DC: US Government Printing Office).

Van Waarden, F. and Lehmbruch, G. (eds.) *Renegotiating the Welfare State* (London: Routledge).

Vieira, P. (2007), 'Ottawa Could Spark Trade War; Action on Single Regulator Likely to Anger Provinces', *The National Post*, 18 October, FP1.

Vogel, D. (1995), *Trading Up: Environmental and Consumer Regulation in a Global Economy* (Cambridge, MA: Harvard University Press).

Vouga, J–P. (1964), 'Le Fédéralisme et la Coopération Intercantonale', *Annuaire Suisse de Science Politique* 4, 83–101.

Wachendorfer–Schmidt, U. (ed.) *Federalism and Political Performance* (London: Routledge).

Wald, M. (2008), 'Environmental Agency Tightens Smog Standards', *The New York Times*, 13 March. <http://www.nytimes.com/2008/03/13/Washington/13enviro.html?hp>

Walter, F. (1990), *Les Suisses et L'environnement: Une Histoire du Rapport à la Nature du XVIIIe Siècle à Nos Jours* (Geneva: Éditions Zoé).

Wasylenko, M. (1991), 'Empirical Evidence on Interregional Business Location Decisions and the Role of Fiscal Incentives in Economic Development', in Herzog, Jr. and Schlottmann (eds.).

—— (1997), 'Taxation and Economic Development: the State of the Economic Literature', *New England Economic Review* (March/April).

Watkins, N. and Burton Jr., J. (1973), 'Employers' Costs of Workmen's Compensation', *Supplemental Studies for the National Commission on State Workmen's Compensation Laws*. vol. II. (Washington DC: Government Printing Office).

Watts, R. and Brown, D. (eds.) (1989), *Canada: The State of the Federation, 1989* (Kingston ON: Institute of Intergovernmental Relations).

Weale, A., Pridham, G., Cini, M. and Konstadakopulos, D. (2000), *Environmental Governance in Europe: An Ever Closer Ecological Union?* (New York: Oxford University Press).

Webb, K. (1988), *Pollution Control in Canada: The Regulatory Approach in the 1980s* (Ottawa: Law Reform Commission of Canada).

Weibust, I. (2003), 'Implementing the Kyoto Protocol: Will Canada Make It?' in Carment et al. (eds.).

—— (2005), 'A (Slow) Burning Issue: Convergence in National Regulation of Dioxins from Incineration', *Policy and Society* 24:2, 46–73.

Weidenbaum, M. (1979), 'The High Cost of Government Regulation', *Challenge* 22:5, 32–9.

—— (1992), 'Return of the "R" Word: the Regulatory Assault on Economy', *Policy Review* 59:Winter, 40–3.

—— (1995), 'A Neglected Aspect of the Global Economy: The International Handicap of Domestic Regulation', *Business Economics* 30:2, 37–40.

Williamson, O. (1975), *Markets and Hierarchies: Analysis and Antitrust Implications* (New York: Free Press).

—— (1989), 'Transaction Cost Economics', in Schmalensee and Willig (eds.).

Winham, G. 1972. 'Attitudes on Pollution and Growth in Hamilton, or "There's an awful lot of talk these days about ecology"', *Canadian Journal of Political Science* 5:3, 389–401.

Winter Jr., R. (1977), 'State Law, Shareholder Protection and the Theory of the Corporation', *Journal of Legal Studies* 6:2, 251–92.

Wristen, K. (1999), *The National Sewage Report Card (Number Two)* (Vancouver: Sierra Legal Defence Fund).

Wurzel, R. (2002), *Environmental Policymaking in Britain, Germany and the European Union: The Europeanisation of Air and Water Pollution Control* (Manchester: Manchester University Press).

Xing, Yuqing (1995), 'International Trade and Environmental Regulations' (Ph.D. dissertation, University of Illinois at Urbana–Champaign).

Zank, N. (1996), *Measuring the Employment Effects of Regulation: Where Did the Jobs Go?* (Westport CT: Quorum Books).

Zax, J. (1988), 'The Effect of Jurisdiction Types and Numbers on Local Public Finance', in Rosen (ed.).

Zimmermann, F. and Wendell, M. (1951), *The Interstate Compact since 1925* (Chicago: Council of State Governments).

Zimmerman, J. (1991), *Federal Preemption: The Silent Revolution* (Ames: Iowa State University Press).

—— (1996), *Interstate Relations: The Neglected Dimension of Federalism* (Westport, CT: Praeger).

—— (1999), *The New England Town Meeting: Democracy in Action* (Westport CT: Praeger).

Zodrow, G. (ed.) (1983), *Local Provision of Public Services: The Tiebout Model after Twenty–Five Years* (New York: Academic Press).

Zürcher, J. (1978), *Umweltschutz als Politikum* (Bern: Francke Verlag).

Zürn, M. and Joerges, C. (eds.) (2005), *Law and Governance in Postnational Europe: Compliance beyond the Nation–State* (Cambridge, UK: Cambridge University Press).

Zürn, M. and Neyer, J. (2005), 'Conclusions–The Conditions of Compliance', in Zürn and Joerges (eds.).

Zwick, D. and Benstock, M. (1971), *Water Wasteland: Ralph Nader's Study Group Report on Water Pollution* (New York: Grossman Publishers).

Zyma, R. (1991), 'Performing Environmentally Sensitive Siting Studies', *Site Selection [Industrial Development Section]* (October), 1082.

Index